Henry Chichester Hart

Some Account of the Fauna and Flora of Sinai, Petra, and Wady 'Arabah

Henry Chichester Hart

Some Account of the Fauna and Flora of Sinai, Petra, and Wady 'Arabah

ISBN/EAN: 9783744795715

Printed in Europe, USA, Canada, Australia, Japan

Cover: Foto ©berggeist007 / pixelio.de

More available books at **www.hansebooks.com**

SOME ACCOUNT OF THE

FAUNA AND FLORA OF SINAI, PETRA,

AND WÂDY 'ARABAH

BY

HENRY CHICHESTER HART
B.A., F.R.G.S., F.L.S.

London
PUBLISHED FOR
THE COMMITTEE OF THE PALESTINE EXPLORATION FUND
BY
ALEXANDER P. WATT
2 PATERNOSTER SQUARE
1891

CONTENTS.

CHAPTER I.
INTRODUCTORY   PAGE 1

CHAPTER II.
'AYÛN MÛSA TO WÂDY LEBWEH   3

CHAPTER III.
WÂDY LEBWEH TO MOUNT SINAI   14

CHAPTER IV.
MOUNT SINAI TO 'AKABAH   21

CHAPTER V.
'AKABAH   27

CHAPTER VI.
'AKABAH TO MOUNT HOR   30

CHAPTER VII.
PETRA AND MOUNT HOR; WÂDIES HARÛN (ABU KOSHEIBEH) AND MÛSA; JEBEL ABU KOSHEIBEH   33

CHAPTER VIII.
WÂDY HARÛN TO THE DEAD SEA   42

CONTENTS.

CHAPTER IX.
SOUTH END OF THE DEAD SEA ... 46

CHAPTER X.
GHÔR ES SAFIEH TO GAZA ... 57

CHAPTER XI.
GAZA TO JAFFA ... 63

CHAPTER XII.
JERUSALEM ... 67

CHAPTER XIII.
JERICHO AND NORTHERN GHÔR ... 70

LIST OF SPECIES.

RANUNCULACEÆ; MENISPERMACEÆ	79
BERBERIDEÆ; PAPAVERACEÆ; FUMARIACEÆ; CRUCIFERÆ	80
CAPPARIDEÆ	82
RESEDACEÆ; CISTINEÆ; SILENEÆ	83
PARONYCHIACEÆ; CERATOPHYLLEÆ; MOLLUGINEÆ; TAMARISCINEÆ	85
HYPERICINEÆ; MALVACEÆ; TILIACEÆ; GERANIACEÆ	86
ZYGOPHYLLEÆ	87
RUTACEÆ; SIMARUBEÆ; TEREBINTHACEÆ	88
LEGUMINOSÆ	89
ROSACEÆ	92
LYTHRARIEÆ; CUCURBITACEÆ; FICOIDEÆ; CRASSULACEÆ	93
UMBELLIFERÆ; RUBIACEÆ	94
DIPSACEÆ; COMPOSITÆ	95
ERICACEÆ; PRIMULACEÆ; OLEACEÆ; SALVADORACEÆ; APOCYNEÆ; ASCLEPIADEÆ	99

CONTENTS. vii

	PAGE
GENTIANACEÆ ; CONVOLVULACEÆ	100
BORAGINEÆ	101
SOLANACEÆ ; SCROPHULARIACEÆ	102
OROBANCHACEÆ ; ACANTHACEÆ	103
GLOBULARIEÆ ; VERBENACEÆ ; LABIATÆ	104
PLUMBAGINEÆ	105
PLANTAGINEÆ ; CYNOCRAMBEÆ ; SALSOLACEÆ	106
AMARANTACEÆ	107
POLYGONEÆ ; NYCTAGINEÆ ; THYMELÆACEÆ	108
LORANTHACEÆ ; EUPHORBIACEÆ	109
URTICACEÆ ; CUPULIFERÆ	110
ARACEÆ ; PALMÆ ; TYPHACEÆ	111
IRIDACEÆ ; AMARYLLIDACEÆ ; COLCHICACEÆ	112
LILIACEÆ	113
ASPARAGACEÆ ; JUNCACEÆ ; CYPERACEÆ	114
GRAMINEÆ	115
CONIFERÆ	117
EQUISETACEÆ ; FILICES ; CHARACEÆ ; MUSCI	118
HEPATICÆ	119
LICHENES	120

AN ANALYSIS OF THE FLORA OF SINAI AND GENERAL REMARKS ON ITS BOTANY, AND THAT OF THE DEAD SEA BASIN.

DESCRIPTION OF SINAI	123
SOURCES OF INFORMATION AND REMARKS ON TABLE OF FLORA OF SINAI	125
TABULAR VIEW OF THE FLORA OF THE SINAITIC PENINSULA	128-144
LIST OF ORDERS ARRANGED ACCORDING TO THE NUMBER OF THEIR SINAITIC SPECIES	145
DESERT FLORA	146
MEDITERRANEAN FLORA OF SINAI ; PLATEAUX OR MONTANE FLORA OF SINAI, AND GENERAL REMARKS	149
SINAITIC SPECIES	154
ON THE FLORA OF THE GHÔR, OR VALLEY OF THE DEAD SEA	156

CONTENTS.

TROPICAL FLORA OF THE DEAD SEA BASIN.

	PAGE
MENISPERMACEÆ ; PARONYCHIEÆ ; MALVACEÆ ; TILIACEÆ ; CAPPARIDEÆ ; SIMARUBEÆ	159
MORINGEÆ ; LEGUMINOSÆ ; CUCURBITACEÆ ; FICOIDEÆ ; COMPOSITÆ	160
SALVADORACEÆ ; ASCLEPIADEÆ ; BORAGINEÆ	161
SOLANACEÆ ; SALSOLACEÆ ; AMARANTACEÆ ; NYCTAGINEÆ ; LORANTHACEÆ ; EUPHORBIACEÆ	162
CYPERACEÆ	163

ADDITIONS TO PALESTINE FLORA.

MENISPERMACEÆ ; CRUCIFERÆ ; RESEDACEÆ ; SILENEÆ	163
PARONYCHIEÆ ; TAMARISCINEÆ ; ZYGOPHYLLEÆ ; LEGUMINOSÆ ; FICOIDEÆ ; RUBIACEÆ	164
DIPSACEÆ ; COMPOSITÆ ; ASCLEPIADEÆ ; GENTIANEÆ ; SCROPHULARIACEÆ	165
LABIATÆ ; PLANTAGINEÆ ; SALSOLACEÆ ; AMARANTACEÆ ; NYCTAGINEÆ	166
THYMELÆACEÆ ; EUPHORBIACEÆ ; SALICINEÆ ; TYPHACEÆ ; LILIACEÆ ; CYPERACEÆ ; GRAMINEÆ ; EQUISETACEÆ	167
CHARACEÆ ; MUSCI	168
HEPATICEÆ	170
LICHENES	172

INSECTA, ETC.

ARACHNIDA	176
MYRIAPODA	177
COLEOPTERA	178
HYMENOPTERA	180
LEPIDOPTERA	181
DIPTERA	182
NEUROPTERA	182
ORTHOPTERA	182
RHYNCHOTA	183

CONTENTS.

MOLLUSCA.

	PAGE
MOLLUSCA FROM 'AKABAH, RED SEA.	
PTEROPODA ; GASTEROPODA	189
CONCHIFERA	193
FROM 'AIN MÛSA, GULF OF SUEZ.	
GASTEROPODA	194
CONCHIFERA	195
FROM THE MEDITERRANEAN AT JAFFA	196
GASTEROPODA	200
CONCHIFERA	201
LAND AND FRESH-WATER MOLLUSCA	202

REPTILIA.

ORDER OPHIDIA	209
ORDER LACERTILIA	210

AVES.

SPECIES OF BIRDS MET WITH IN THE WINTER MONTHS	215
SPECIES OF BIRDS WHICH VISIT THE PENINSULA	227

MAMMALIA.

SPECIES OF ANIMALS, ETC., MET WITH IN THE PENINSULA, WÂDY 'ARABAH, AND SOUTHERN PALESTINE	233-238

CONTENTS.

MAPS.

ROUTE MAP *frontispiece*
GEOLOGICAL MAP *to face p.* 79

PLATES.

PL. I.—1. GALIUM PETRÆ; 2. DAPHNE LINEARIFOLIA	*to face p.*	95
,, II.—IPHIONA SCABRA	,,	96
,, III.—GOMPHOCARPUS SINAICUS	,,	100
,, IV.—BOUCEROSIA AARONIS	,,	100
,, V.—LINARIA FLORIBUNDA	,,	102
,, VI.—LINDENBERGIA SINAICA	,,	103
,, VII.—LORANTHUS ACACIÆ	,,	109
,, VIII.—XIPHION PALÆSTINUM	,,	112
,, IX.—PANCRATIUM SICKEMBERGERI	,,	112
,, X.—INSECTA	,,	175

ERRATA.

Page 95, *delete* '[Plate XVI., fig. 1].'
,, 100, *for* 'aavonis,' *read* 'Aaronis.'
,, 100, *delete* '[Plate XVII., figs. 1 to 8].'
,, 108. For a figure of 'Daphne,' see opposite p. 95, *ante*.
,, 109, *delete* '[Plate XVI., fig. 2].'
,, 119, *for* 'Encalypta,' *read* 'Eucalypta.'
,, 201, *for* 'Mediterranean at Jaffa,' *read* 'Red Sea.'
,, 202. The remark at the head of this page was written in 1884, and is now probably incorrect.

SOME ACCOUNT OF THE

FAUNA AND FLORA OF SINAI, PETRA, AND WÂDY 'ARABAH.

CHAPTER I.

INTRODUCTORY.

EARLY in the summer of 1883 my friend Professor Hull, Director of the Geological Survey of Ireland, proposed to me that I should accompany him as a volunteer on a geological and surveying expedition to Sinai and the Dead Sea, of which he was about to take the leadership under the auspices of the Palestine Exploration Society.

With the main object of studying the botany of this region, and as far as possible also other branches of its natural history, I accepted this friendly offer. I was chiefly induced to do so by the assurance I received from Professor Oliver, of Kew, that, whatever our Continental brethren may have accomplished, few British botanists had as yet turned their attention to Sinai. He at the same time promised his valuable assistance in the determination of my specimens upon my return — a promise since fulfilled in a manner which entitles him to my sincerest thanks. Another welcome consideration which helped to determine me was that of a grant of money from the Scientific Fund of the Royal Irish Academy.

I feel bound to take this earliest opportunity of expressing my grateful sense of the courtesy of the Rev. Canon Tristram, the well-known authority on the Natural History of Palestine, who has helped me with his advice before starting, and his scientific knowledge since my return.

To him the determination of my species of birds, as well as of land and freshwater molluscs, is almost entirely due, and his recent work on the 'Fauna and Flora of Western Palestine' has been continually consulted in preparing the present account.

To Dr. Gunther, F.R.S., and to Messrs. Waterhouse and Thomas, of the British Museum, my thanks are due for the naming of other smaller collections of mammals, reptiles, and beetles. Mr. Edgar Smith, of the conchological department, has also been good enough to render me as much assistance as his duties would permit, in searching for information on the mollusc-fauna of the Red Sea.

To Mons. Edmond Boissier, the eminent Swiss botanist and author of the invaluable 'Flora Orientalis,' I desire to tender my warmest acknowledgments. He has very kindly determined for me some of the more intricate genera, which his unrivalled knowledge and extensive Oriental herbarium enable him to deal with satisfactorily. Of Mons. Boissier's 'Flora Orientalis' I have constantly availed myself in dealing with the flora of Sinai. Botanists whose inclinations turn, as mine do, to the geographical distribution of plants, will find this work, which is now complete, a perfect storehouse of information.

Reference must here be made to the 'Ordnance Survey of Sinai,' published in 1869, where much valuable information on the physical features and natural history of the Peninsula will be found, especially in the appendices of Mr. Wyatt. An interesting paper by Mr. Lowne, on the Flora of Sinai, in the Journal of the Linnean Society for 1865, may also be referred to; his nomenclature, however, differs widely from that at present adopted. There is little other botanical literature available; Decaisne's 'Florula Sinaica,' published in the *Annales des Sciences Naturelles* in 1836, in which many new species are described, is difficult to obtain separately; it is, however, very valuable, but the collections of Schimper and others, distributed throughout the herbaria of Europe, and duly recorded in Boissier's 'Flora Orientalis,' have nearly doubled Decaisne's original total.*

I must not omit to acknowledge the judicious and kindly guidance by which (with the assistance of our most efficient interpreter and conductor,

* Since writing the above, Europe has lost one of her most famous botanists. Mons. Boissier died in September, 1885.

Bernard Heilpern) Professor Hull brought our travels to a safe conclusion. In a volume recently published by the Society, he has given the public an account of our experiences, and to it, and its Appendix by Major Kitchener, the reader may turn for fuller geological, geographical, and other information relative to our explorations. To the other members of our party, for their assiduity in obtaining specimens for me, I shall feel for ever grateful.

In these pages, which owe their appearance to the liberality of the same Society, I propose in the first place to give a running account of the collections made in the order in which they were gathered, with such extracts from my journal as may serve to illustrate them. Afterwards I will enumerate in detail the various species which I have identified, and conclude with an endeavour to give a full account and analysis of the Flora of Sinai, or rather of the Sinaitic peninsula of Arabia Petræa.

By the kind permission of the Royal Irish Academy, the systematic list of plants is reproduced here from their Transactions. The specimens themselves are in the Herbaria of Kew and the British Museum.

CHAPTER II.

'AYÛN MÛSA TO WÂDY LEBWEH.

HAVING left Suez on Saturday, November 10, 1883, we took up our quarters till Monday at 'Ayûn Mûsa, the usual starting-place for Sinai. A description of the gardens here, with the introduced plants found about them, has been given by Mons. Barbey, in his recent volume ' Herborisations au Levant,' who visited them at a more auspicious season. His tour did not elsewhere cover the ground we visited till reaching Bîr es Sebâ.

At 'Ayûn Mûsa my hopes fell to a low ebb. With the exception of a couple of showy flowering shrubs (*Lantana camera, Linn*, and *Cassia bicapsularis, Linn.*) in the gardens of date palm, bounded by prickly pear, there appeared to be hardly a vestige of unwithered vegetable life. Closer inspection, however, yielded dead flowers and ripe seed capsules of several species, all of which were carefully preserved for comparison with subsequent gatherings. One species, *Ceratophyllum demersum, L.*, found drifting in the gulf, and probably derived from the canal, was not met with again. A prostrate prickly grass in the sandy stony flat between the wells ('Ayûn Mûsa) of Moses and the gulf has been named for me by Mons. Boissier, *Sporobolus spicatus, Vahl.*

In these enclosures, and around their edges, were bushes of tamarisks and 'ghurkûd,' *Tamarix nilotica, Ehr., T. articulata, Vahl.* (?), and *Nitraria tridentata, Desf.* The latter is a prickly, fleshy-leaved shrub with small orange berries, greedily eaten by camels. It belongs to the 'bean-caper' family (Zygophyllaceæ), well represented in the desert.

From one of the wells numerous univalves, all of one species, *Melania tuberculata, Mull.*, were obtained. The net produced nothing else except the larvæ of a gnat. A chamæleon (*Chamæleo vulgaris, Linn.*) and a small very nimble brown lizard (*Eremias gutto-lineata*) were captured close by. The former was pointed out to me by a Bedawîn on a stunted palm-tree, else I should assuredly have passed it by, so closely did it resemble the branch along which it clung.

The chief attraction at this oasis was in the birds, of which several species were obtained. Amongst these were the white wagtail and the willow-wren (*Motacilla alba, Linn.*), and *Phylloscopus rufus, Bechst.*). A buff-backed heron, *Ardeola russata, Wagl.*, was seen but not shot: this is the bird which does duty for the 'white ibis' amongst visitors. A little cock-tailed warbler with a song and habit of a wren, *Drymoeca inquieta, Rupp.*, as well as the blue-throated robin, *Cyanecula cæruleculus, Pall.* (the one with the entirely blue throat), was shot here.

Across the sand to the shores of the gulf many kinds of sea shell were gathered. A detailed account of these, as well as of those obtained at 'Akabah, will be given later on. Few specimens worth preserving were met with, but they were for the most part identifiable. At the water's edge a stork gave me a long shot, and several dunlins were flying about.

At evening the air was filled with the attractive notes of species of cicada, and the quaint call of an owl (*Athene meridionalis, Risso.*), the 'boomey' of the Arabs, was for the first time heard.

Insect life was almost suspended, but a few small beetles (*Adesmia, Acis*), ants (*Camponotus*), and a spider or two, as well as a torpid scorpion, were captured about here, and between this and Wâdy Nasb.

Excepting at wells, met with at rare intervals, life of all kinds was very scarce in this lower desert portion of Sinai. The appearance of a bird within a quarter of a mile in these wastes was a signal for a general call to arms amongst the gunners, and the gurgling sound of the Bedawîn camel-driver summoned his obstinate beast to kneel and let his rider dismount and stalk a distant Egyptian vulture or a raven. These two birds, *Neophron percnopterus, Linn.*, and *Corvus umbrinus, Hedenb.*, were frequently in sight, but rarely in range.

After a day or two, when my Bedawîn lad, Khalîl, had discovered which of us two was master, I generally travelled on foot, letting my camel-driver keep me in view till wanted. For this interesting and faithful son of the desert I conceived a great liking. This feeling towards the Arabs is very frequently indulged in by inexperienced travellers in the East.

As fast as I made gatherings, I was able to deposit them on the back of my admirable beast of burthen. For this purpose I had two sets of camel bags and drying boards, as well as multifarious swinging gear; guns, spy-glass, water-bottle, shoulder-bag, spirit-cylinder, portfolios, insect-box, *et hoc genus omne*.

The country traversed was of gravel and sand, with occasional outcrops of limestone. This limestone sand is sometimes finely and regularly granulated, as near Wâdy Sudur, a condition not observed by us in other parts of Sinai. The view of the Jebel Rahah mountains across the Gulf of Suez was superb.

Our direction lay nearly parallel to this arm of the Red Sea, gradually widening the distance between us and the coast-line. The sky was of a brilliant blue, and the temperature rarely hot enough to make walking disagreeable. The following plants were observed in Wâdy Sudur : *Zilla myagroides, Desf.; Retama retam., Forsk.; Alhagi maurorum, D.C.; Acacia Seyal, Del.; Deverra tortuosa, Gærtn ; Anabasis articulata, Forsk.; Reaumuria vermicularis, Linn. (R. palæstina, Boiss.); Fagonia cretica, Linn., var. glutinosa et vars.; Erodium glaucophyllum, Ait.; Citrullus colocynthis, Lehr.; Artemisia iudaica, Linn.; Odontospermum graveolens*, S. *Bip.*; *Gymnocarpus fruticosus*, Pers.; *Paronychia desertorum*, Boiss.; *Ærua javanica*, Juss.; *Heliotropium luteum*, Poir.; *Aristida obtusa*, *Del*. Most of these are strictly desert species of continual occurrence in the lower parts of the peninsula, and will seldom again be referred to. In Wâdy Sudur *Farsetia Ægyptiaca*, *Turr.*, and *Anabasis setifera*, *Moq*., were also obtained.

The Citrullus bore its ripe fruit, orange-coloured and about the size of a billiard-ball, trailing on the gravel and sand in many places.* The felted Ærua was laden with tassels of wool, the remains of its withered inflorescence ; the variety, with narrower leaves and more rigid habit, occurred later on. *Acacia Seyal* was a revelation of spinousness whose branches even the camel can only nibble with care. It is a low flat-topped bush, often only 4 or 5 feet high, but with a trunk of considerable thickness.

A Matthiola, probably *M. arabica*, *Boiss*., occurred, and a large cabbage-leaved sticky Hyoscyamus, *H. muticus*, *Linn*., with showy yellow and purple-veined flowers, was pointed out to me as the 'Sekkaran,' which the Arabs are said to inhale in their narghilis as an intoxicant.

The pretty little woolly Reaumuria, with its densely imbricated leaves, was, after much searching, found in blow at last. A wiry, nearly leafless Deverra was in full flower and seed, with a strong but not unpleasant smell of fennel.

The marked characteristics of these desert plants soon become familiar. They have usually a whitened appearance, which was perhaps somewhat heightened at the season of my visit. This is due to woolliness, or scaliness, or some other colouring integument, and is frequently accompanied by heavy odours, succulent or glaucous foliage. Spines, prickles, hooked or clinging hairs are also characteristic, and the whole plant is not unfrequently found to be steeped in a strong viscid exuda-

---

* The Arabs use this species (the colocynth) as a purgative. A fruit is split into halves, the seeds scooped out, and the two cavities filled with milk ; after allowing it to stand for some time, the liquid, which has absorbed some of the active principle of the plant, is drunk off. I refer my readers for further valuable information of this nature to an article in the *British Medical Journal* of April 11, 1885, by my friend and companion, Dr. Gordon Hull. I trust he will forgive me for correcting an error into which I unfortunately led him. The plant which he speaks of ' with short succulent jointed segments ' as being very common and used for sore eyes is not Zygophyllum, but *Anabasis* (*Salsola*) *articulata*.

tion. Noteworthy instances of the above peculiarities will be given farther on.

Of the Sinaitic mountains, no part was as yet visible; we were, however, gradually rising above sea-level, and with the cooler atmosphere there was a steady increase also in the quantity of vegetation. A very fragrant bushy Artemisia, *A. santolina, Linn.*, had become frequent, and is subsequently one of the most characteristic plants of the flat wâdies.

In Wâdy Sudur *Cleome arabica, Linn.; Pennisetum dichotomum, Del.*, and *Elionurus hirsutus, Vahl.*, were secured in good condition, except the latter grass, which is so closely eaten by camels that it is hard to obtain good specimens.

Anabasis articulata, Forsk., is a prevalent low-sized species; its dried twigs are always topped by a few scales, the remains of the floral envelopes. These are occasionally a showy red or claret colour, and give a brilliant effect, sometimes equalling that of red heather at a distance. It is perhaps the commonest species throughout Sinai; *Gymnocarpum fruticosus, Forsk.*, however, is nearly as abundant. The Anabasis, whose slenderer twigs are, I believe, all lost and withered at this season, accumulates round its roots blown hillocks of sand a couple of feet high, favourite hiding places for lizards, and burrowing ground for ants and the smaller rodents. The Bedawin called this plant ' Erimth.'

The vegetation is scattered in tufts amongst the sand and gravel; except in the occasionally moistened wâdy beds these tufts are usually isolated and often far apart.

On the 13th, at about 350 feet above sea-level, we entered a bed of chalk intermixed with white marls strewed with chert, fossils, and selenite. We reached Ghurundel by moonlight. Tamarisks and palms (*Tamarix nilotica, Pall.; Phœnix dactylifera, Linn.*) form here a pleasant grove; Zilla, Nitraria, and most of the species above mentioned, are plentiful.

At Wâdy Ghurundel (' Elim ')* I obtained some fresh species of birds.

* This wâdy must not be confounded with others of the same name in Sinai and Edom. A notable instance of confusion occurs in the ninth chapter of the English translation of Laborde's 'Arabia Petræa,' 1836, where the translator quotes several pages of description of the present wâdy from Burckhardt, to illustrate Laborde's short and correct mention of Wâdy Ghurundel, near Petra.

Of these *Saxicola isabellina, Rupp.* (Menetries' Wheatear) was several times seen and shot. The 'Persian lark' (*Certhilauda alaudipes, Desf.*) and the striolated bunting (*Emberiza striolata, Licht.*) were obtained, only single specimens being as yet seen and secured of each. Ravens and willow-wrens tenanted this wâdy.

The first large quadruped's tracks were pointed out by the Arabs; they exclaimed 'dhaba'—that is to say, 'hyæna.'

Another lizard, *Agama ruderata, Riv.*, and a skink, *Sphænops capistratus, Wagl.*, were captured here. The latter I found on kicking to pieces an ant hill, the home of a species of Camponotus, *C. pubescens*. The lizard was afterwards very common throughout Sinai to the Dead Sea. He was easy to catch, and his comical habit of standing at bay with his tail cocked and his disproportionately large jaws wide open was instructive; no doubt it terrified troops of smaller foes. Like most true natives of the desert he was sand-coloured, though the tail had some dull blackish rings. Another lizard, *Eremias guttata*, was most difficult to catch; by pelting him with handfuls of sand, which confuses and stops his movements for an instant, combined with a sudden rush, it may be done.

The rock here is a white cretaceous limestone. The bed of the wâdy is cut deeply into marly deposits, leaving sheer mud-banks sometimes 8 feet high. The bed of this periodic stream was now perfectly dry. From the appearance of these deposits, and those in other places, Professor Hull considered there was evidence of a much greater rainfall in recent times.

On the tamarisk branches a curious buff-coloured chrysalis-like appendage was frequently observed. It was about the consistency of tough paper, half an inch long, but more brittle, and proved to be the egg case of a species of Mantis. A large black beetle, *Prionotheca coronata, Oliv.*, was the only large insect found in Wâdy Ghurundel.

Several plants were here first met with; the most conspicuous was a shrubby mignonette, *Ochradenus baccatus, Del.*, thenceforward characteristic of the lower desert wâdies, and sometimes, where protected by acacia trees from camels, 6 or 8 feet high.

Here or nearer to Wâdy Useit, I noticed for the first time a second species of acacia, *A. tortilis, Hayne*, less spiny and usually larger and more upright than *A. Seyal, L.* I met only these two acacias in the peninsula, but I found a third and much finer one (*A. læta, R. Br.*) at the south end of the Dead Sea. *A. nilotica, Del.*, also occurs in Sinai. *A. tortilis* is commoner in the 'Arabah than elsewhere.

Other species were: *Cucumis prophetarum, Linn.; Polycarpæa fragilis, Del.; P. prostrata, Dcne.; Zygophyllum album, Linn.; Fagonia cretica, Linn.*, var. *arabica; Lithospermum callosum, Linn.; Cressa cretica, Linn.; Euphorbia cornuta, Pers.; Juncus maritimus, Linn.*, β *arabicus; Typha angustata, B. et C.; Cynodon dactylon, Pers.; Phragmites communis, Linn.*, var. *gigantea*. This latter species, which reaches a height of 10 or 12 feet with its erect plume of florescence, is a truly handsome grass. It appears to have frequently done duty for *Arundo Donax, L.*, in Sinai.

Many withered Chenopods occurred here, the identifiable species being *Suæda vermiculata, Forsk.; Atriplex leucoclada, Boiss., A. halimus, Linn; Anabasis setifera, Moq.;* and *A. (Salsola) articulata, Forsk.* At Wâdy Useit occurred a little grove of date palms, some of them at least 40 feet high. There is only one other species, the doum palm (*Hyphæne thebaica, Del.*), in Sinai. It occurs near 'Akabah and at Tôr.

From about Wâdy Sa'al small burrows, from the size of a small rabbit-hole to the little perforation of a species of ant, *Camponotus compressa, Fab.*, become numerous. These belong chiefly to species of Acomys, Gerbillus, and Psammomys, but it was some time before I succeeded in capturing any of these animals. On several occasions I saw individuals of the Gerbille genus of sand-rats. These animals usually burrowed in the sand-hills accumulated about the stumps of anabasis and tamarisk; their abundance here was as nothing compared with their numbers in the Wâdy 'Arabah later on. Jerboas were not seen in Sinai.

At night in the dinner tent our lights usually attracted a few nocturnal insects, which I captured from time to time.

A hornet, *Vespa orientalis, Linn.*, was the only insect frequently to be seen in the day-time. Nature rests herself in the desert almost as thoroughly as in an Arctic winter; in the latter case she sleeps during an excessive cold, in the former she exhausts her strength during an extreme heat. Nevertheless many late flowering plants still occasionally held their petals, and it was not many days ere we gathered the first harbingers of spring. Possibly these latter should be called hybernal. A few species, as *Cleome arabica, Linn.*, are in their prime at present for examination, being in full flower and fruit. This Cleome is one of the most viscid plants met with, taking many weeks to dry, and never shaking off the adhering sand. It has small deep purple flowers and longish pods.

A black snake, probably *Zamenis atrovirens, Shaw.*, var. *carbonarius*, was killed here, but I was informed it was last seen with the cook. Whether it subsequently passed under examination in the dinner tent I cannot say, but I never succeeded in identifying it.

Desert larks representing three genera were obtained; one of these, Certhilauda, has been already mentioned. Other two, *Ammomanes deserti, Licht.*, and *Alauda isabellina, Bonap.*, were also shot. The latter is one of the most frequently met with of the true inhabitants of the desert. The Persian lark (*Certhilauda desertorum, Rupt.*), a bird about the size of our song-thrush, has a low sweet song, uttered while on the ground, and not much stronger than or unlike our robin's winter warble. A large and handsome black and white chat (*Saxicola monacha, Temn.*) was shot in Wâdy Hamr. Tracks of gazelles were here first observed.

At Wâdy Hamr we are crossing beds of a highly coloured red sandstone, which has replaced the white and black weathered limestone. The black and white chats are more conspicuous amongst these rocks; when at rest on a chalky surface dotted with fragments of chert these birds are not quickly seen. The desert larks are, however, the most securely assimilated to the soil. The females of some chats (*e.g., S. monacha*) are more protectively coloured than the males.

The sandstone which we were now traversing is the regular inscription rock of the desert, on which the Bedawîn of all ages have delighted to air their calligraphy, and not unfrequently impose upon travellers with their rude tribe-marks.

Our direction was mainly south-east, and steadily rising. At the head of Wâdy Hamr, about 1,300 feet above the sea-level, we obtained our first view of the Sinaitic mountains. Jebel Serbâl stood out, grand and rugged, straight ahead of us, looking about one-half of his real distance from us, so excessively clear was the atmosphere.

Leyssera capillifolia, D.C., was gathered here for the first time, and the favourite camel grass, *Elionurus (Cœlorachis) hirsuta, Vahl.*, was gathered in flower.

FAUNA AND FLORA OF SINAI, PETRA, AND WADY 'ARABAH. 11

Having left Wâdy Hamr, and crossed Sârbût el Jemel at a height of about 1,700 feet above sea-level, we came out on a wide sandy plain, Debbet er Ramleh, lying about 1,700 to 1,850 feet above sea-level. This is the largest expanse of sand in Sinai, and covers about thirty square miles. Some very interesting species were gathered here. The two species of Polycarpæa already mentioned, with the Cleome, abound.

Seetzenia orientalis, Dcne.; Glinus lotoides, Linn. (not in flower); *Monsonia nivea, Dcne.; Pancratium Sickembergeri, A. et S.; Danthonia Forskahlii, Linn.; Aristida plumosa, Linn.;* and *A. obtusa, Del.* These were all obtained in flower, and the white and perfect Pancratium was at its best. It is a lovely flower, and I secured many bulbs here and elsewhere. No leaves were yet in sight, but in some cases the petals had fallen, and the seed pod was filling, showing that the leaves are certainly not synanthous, though appearing soon after the flowers. Plants of this species subsequently growing with me did not exhibit the remarkable twisting described as characterizing their leaves. On this Pancratium, which was first discovered by Sickemberger near Cairo, some interesting remarks will be found in Barbey's 'Herborisations,' already mentioned.*

The Aristidæ, small glaucous grasses with long feathery awns, are amongst the prettiest of desert forms.

At a lower level near this, *Lycium europæum, Linn.*, was plentiful, and in full flower. It is visited by a small copper butterfly, the first of its family met with, which is poorly represented in this dry region. Formicidæ and Acridiidæ (ants and locusts) are perhaps the most abundant insects.

In Wâdy Nasb several fresh species occurred. Unrecognisable fragments awoke my regrets at the season selected from time to time.

The following were determined: *Morettia canescens, Boiss.; Astragalus sieberii, D.C.; A. trigonus* (?), *D.C.; Crotalaria ægyptiaca, Bth.;* and *Convolvulus lanatus, Vahl.*

These Astragals were quite withered, and simply well-rooted bunches of strong sharp spines, 2 to 3 inches long, set closely round a stumpy stem; the spines being the hardened woody mid-rib of the pinnate leaves. The only evidence of their past condition lay in the slight cicatrices in the spines marking the points of attachment of the fallen leaf-pinnæ. Of the

* 'Herborisations au Levant,' par C. et W. Barbey (Lausanne, G. Bridel, 1882).

convolvulus, a handsome, erect, shrubby, felted species, with good-sized reddish-purple petals, I obtained a couple of flowers.

Desert partridges were first heard here, but not yet obtained. Chats and larks appeared to be pairing. A shrike, *Lanius fallax, Finsch.*, was first seen and shot. Afterwards this became a familiar species. The ' desert blackstart,' *Cercomela melanura, Temn.*, another very characteristic and prevalent bird of Sinai, was also first met with and obtained here. The chats were *Saxicola leucopygia, Brehm.*, and Menetries' wheatear already mentioned. The trumpeter bullfinch, *Erythrospiza githaginea, Licht.*, was shot here for me by Dr. Hull, who, as well as Mr. Reginald Laurence, brought me specimens from time to time.

In Wâdy Nasb there is a well, and quite a goodly show of acacias, chiefly of the species *A. tortilis, Hayne.*, which was in flower sometimes, and usually in leaf. The leaf segments of this species are larger and fewer in number than in *A. Seyal, L.*, the pods are twisted, and the tree attains a greater size. When old it is less and less spiny, while the reverse seems to be the case in *A. Seyal.*

In this wâdy I gathered *Malva rotundifolia, Linn.*, and *Amarantus sylvestris, Desf.*, by the well, both probably of human origin. The former is cooked and eaten by the Bedawin. *Lycium europæum* has flowers either white or pinkish-purple. Other species met here first were: *Dæmia cordata, Br.*; *Echiochilon fruticosum, Desf.*; *Lavandula coronopifolia, Poir.*; *Crozophora obliqua, Vahl.* (a perennial form of *C. verbascifolia, Juss.* ?); and *Zizyphus spina-christi, W.* The latter was not native, and occurred in a miserable little enclosure by a Bedawin hut at the well. It was less thorny than the native species afterwards gathered, and the fruit somewhat larger, but Mr. Oliver refers it to the same plant, no doubt slightly altered and improved by a rough system of cultivation.

As we are gradually increasing our elevation amongst the wâdies derived from the precipitous escarpment of the Tîh plateau (4,000 to 5,000 feet), so there are more remains of last summer's vegetation—later in flowering, perhaps, and less scorched than the same species below.

Soon after leaving Wâdy Nasb we entered on plutonic formations, a red porphyritic granite, which was thenceforth to accompany us upwards over a large extent of country. The increased quantity of acacias since we left the limestone, and especially on the granite, is noteworthy. Perhaps its ferocious spines require an admixture of silicon.

A locust and a cricket were taken in semi-torpid condition. Scorpions similarly harmless were caught from time to time.

A larger species of lizard, with a handsome blue throat and pectoral, was captured, *Agama sinaitica, Heyden*. The bright colour was all below, and was no reproach upon the perfect assimilation of its upper parts with the desert sandy hues. This lizard hid himself amongst stones, and it was with difficulty I dislodged him from a hole which he filled with his body and fortified with his distended and savage little jaws.

Having crossed a high ridge of granite, Râs Sûwig, at about 2,400 feet above sea-level, from whence Jebel Serbâl looked magnificent, we descended into a wâdy which yielded several new plants. *Pancratium Sickembergeri, A. et C.*, was found in flower here also. A small bulb, apparently an Allium, was brought to me by some Bedawîn. It flowered under Mr. Burbidge's care at the College Botanic Gardens, and proved to be *A. sinaiticum, Dcne*. These two bulbs and a Uropetalum (*U. erythræum, Debb.*) are, I believe, the only ones which support life in this desert. A few others occur, but at sufficient heights, usually very considerable, to bring them into a different zone of plant life.

At the height of 2,200 to 2,400 feet above sea-level the following species appeared: *Iphiona juniperifolia, Coss.; Sonchus spinosus, Del.;* and a very fetid species, *Ruta tuberculata, Forsk.*, was here first obtained with its yellow flowers.

Major Kitchener brought me branches here of the first Capparis I had seen, *C. galeata, Fresen*.

Lichens of two species at least occurred, one on the bark of acacia, and the other on sandstone.

In Wâdy Khamîleh desert partridges, *Caccabis Heyi, Temn.*, were frequent, and some were shot. Two desert plants occurred in some quantity, *Lotononis Leobordea, Linn.*, and *Pulicaria undulata, D.C.*

CHAPTER III.

WÂDY LEBWEH TO MOUNT SINAI.

STILL ascending gradually, up Wâdy Lebweh, from 2,500 to 3,500 feet, many interesting Sinai plants were gathered. Most of these are true desert species, which reach about thus far, but they are mixed with others of an intermediate elevation about corresponding to the Mediterranean flora. From here come *Glaucium arabicum, Fres.; Caylusea canescens, St. Hil.; Cleome trinervia, Fres.; Fagonia myriacantha, Boiss.; Tribulus terrestris, Linn.; Peganum harmala, Linn.; Neurada procumbens, Linn.; Santolina fragrantissima, Forsk.; Artemisia herba-alba, Asso.;* et var. *laxiflora, Sieb.; Anarrhinum pubescens, Fres.; Trichodesma africana, R. Br.; Heliotropium undulatum, Vahl.; Gomphocarpus sinaicus, Boiss.; Ballota undulata, Fres.; Teucrium polium, Linn.,* β *sinaicum; Stachys affinis, Fres.; Primula boveana, Dcne.; Acanthodium spicatum, Del.; Forskahlea tenacissima, Linn.; Andrachne aspera, Spr.; Asphodelus fistulosus, Linn.;* and others, the specimens too bad to name. The labiates in the above group are characteristic of the middle and upper zones of Sinai.

On the summit of Zibb el Baheir, at 3,890 feet, a point which all travellers should climb for the sake of the really splendid view, *Gypsophila rokejeka, Del.; Helianthemum Lippii, Pers.; Iphiona montana;* and a Poa, *P. sinaica, St.* (?), were gathered. A Psoralea occurs here also, not found in a recognisable state. It may have been *P. plicata, Del.*

Of the plants just enumerated several are peculiar to Sinai. Others, believed endemic, I found later on Mount Hor in Edom.

In addition to the above it is to be remembered that the majority of the earlier species met with occur throughout. The chief failures are *Cleome arabica, Linn.*, and Salsolaceæ (except Anabasis), which are mostly confined to the lower plain. The variable but always pretty little Fagonia is continually arresting the attention by some new deviation. Sometimes it is glabrous, sometimes viscid, sometimes very leafy; at others a bunch of twigs or thorns, trailing or sub-erect, while the flowers vary much in size. In one form or another it is a very widespread desert form which has received a number of segregational names. The abnormal *Neurada procumbens*, with its curious flat prickly-edged capsule nearly an inch in diameter, was in good condition, but scarce. *Gomphocarpus* was in full flower and fruit; like *Dæmia cordata*, already gathered, and now common, it has a sticky, staining, milky juice, very poisonous according to the Bedawîn. These two Asclepiads, and about five others occurring in Sinai, point to the tropical element in its flora. *Artemisia herba-alba, Asso.*, in several well-marked forms, is henceforth one of the most abundant and highly aromatic plants.

From Zibb el Baheir, which I ascended with Dr. Hull on Sunday, November 16, we had a grand view of the whole mass of Jebel Mûsa (Mount Sinai) and Jebel Katharîna on the south-east, and of Serbâl nearer us to the southward. Down Wâdy Berah the foregoing labiates and composites were prevalent in many places. A little further on is a continuous grove of retem bushes, the first bit of almost luxuriant though limited vegetation I had seen except close to the wells. This wâdy, like most others, is flat, and about half a mile wide, with a slight channel wandering from side to side, and marked by a line of gray-green growth, no doubt fresh and delightful after the rain, which was almost due.

Hares were seen once or twice. I saw one here first, a very long-eared and long-legged whitey-gray animal with a little body (*Lepus sinaiticus, Hemp. et Ehr.*). He was a perfect fiend to travel; no animal ever got out of my sight so quickly. The little southern owl hovered around our camp one or two evenings. A splendid pair of griffon vultures afforded a nearer view here than elsewhere. The Egyptian species is more approachable. Crows and ravens (*C. corax* and *C. umbrinus*) were also tamer in this less frequently traversed route. Indeed, the large birds generally seemed fully aware of the harmless nature of Cairo powder. The lark, *Alauda isabellina, Bon.*, was the commonest of the smaller species. White wagtails, *Motacilla alba, Linn.*, were also very frequent, continually hopping about our tents and camels, quite fearless of man.

The two lizards of the *Agama* genus already mentioned, especially the smaller (*A. ruderata*), were common. I kept some of these alive as far as to Constantinople three months later, but the cold weather there killed the last of them.

The mountains are of red porphyry intersected by numerous dykes of trap. This is surely the proper country for a geologist to come to; no annoying mantles of soil or vegetat'on conceal the rock masses; all is bare and clear, and a good view reveals as much as a shire full of well-borings and railway-cuttings.

The temperature became much colder, falling to within five or six degrees of freezing-point at night, and we found it difficult to keep warm enough in our tents.

Acacia bushes become rare or absent at about 3,500 feet elevation. Acacias may be said to mark the vertical limits of the desert flora, as the date-palm does its horizontal geographical distribution. The desert plants which exceed this range upwards will be found to be mostly Mesopotamian or Syrian species, and not confined to that belt which extends from the Cape Verdes to Scinde.

In Wâdy es Sheikh some large tamarisk-bushes (*T. nilotica*) occur, about 15 feet in height. This plant has about the same upward limit as that of the acacia. On these tamarisks were two butterflies, one of which, *Pyrameis cardui, Linn.*, was obtained; the other appeared to be a fritillary (Argynnys).

The Wâdy es Sheikh is of considerable length, upwards of twenty miles, running east at first, and then south to the base of the Jebel Mûsa group. It lies high, 3,000 to 4,000 feet, and the chief plants in it are Artemisiæ, Santolina, and Zilla, except on the northern sides at the base of whatever shelter from the sun there may be. Here most of the plants lately enumerated occurred still. Some appeared which were less common, as *Zygophyllum album, Linn.; Nitraria tridentata, Desf.; Alhagi maurorum, D.C.; Crozophora obliqua, Vahl.; Pancratium Sickembergeri, A. et S.;* and the labiates and composites of Wâdies Lebweh and Berah. *Gomphocarpus sinaicus, Boiss.*, often arrested attention, shedding its beautifully silky tufts of hair, ready to whisk the attached seeds about the peninsular plains with every breath that blows. *Phagnalon nitidum, Fres.; Anabasis setifera, Moq.;* and *Atriplex leucoclada, Boiss.*, occurred in Wâdy Solâf, so that the Salsolaceæ only require favourable circumstances to appear in the upper country. In Wâdy Solâf, a smaller arm of the Wâdy Sheikh, remarkable sections of marl deposits, many feet in thickness, were examined. These no doubt represent the bed of a large lake of the recent period cut through by streams which once contained a steady supply. Examination of evidence of this nature will form an interesting portion of Professor Hull's results.

At Jebel Watiyeh a fine granite pass connects the eastern and southern prolongation of Wâdy Sheikh. The summit of this I estimated at 4,150 feet above sea-level. On it I obtained *Dianthus sinaicus, Boiss.*; *Buffonia multiceps, Dcne.*; *Arenaria graveolens, Schreb.*; *Cratægus sinaica, Boiss.*; *Cotyledon umbilicus* (?), *Linn.*; *Poa sinaica* (?), *St.*; and most of the species of Zibb el Baheir. The withered *Psoralea* (sp. ?) occurred also. The first two of these are peculiar to Sinai. There was a well-marked difference here in the floras of the north and south side of the peak, the Cotyledon and grass occurring only on the north side, while the Artemisiæ, Anabasis, and other ubiquitous desert species prevailed on the other or southern face.

Laurence caught for me on this crag a locust (*Tryxalis unguiculata, Linn.*), resembling exactly the withered straw-coloured twigs and sand in which he lived.

Further towards Wâdy Suweirlyeh grow *Pyrethrum santalinoides, D.C.*; *Centaurea eryngoides, Lam.*; *Alkanna orientalis, Boiss.*; *Lithospermum tenuiflorum, Linn.*; *Suæda monoica, Forsk.*; *Piptatherum multiflorum, Beauv.*; and of rarer kinds, *Echinops glaberrimus, D.C.*; *Iphiona montana, Vahl.*; *I. juniperifolia, Coss.*; *Anarrhinum pubescens, Fres.*; *Primula Boveana, Dcne.*; and *Teucrium sinaicum, Boiss.*

It was interesting to notice a form of *Cotyledon umbilicus, Linn.*, the only apparently native British dicotyledon I met with in Sinai. It has been gathered here previously by Bové, according to Decaisne, who recorded it under the present name. Unfortunately my specimens are in too bad a condition to determine, consisting only of young leaves and a withered stem. The root was tuberous. It is plentiful on Mount Hor, and is not unlikely to be identical with the new form Dr. Schweinfurth gathered on mountains between the Red Sea and the Nile Valley.[*]

Retama Retem., Forsk., is very common in these high-lying wâdies. It quite takes the place of acacia, and was now laden with its one-seeded

[*] Barbey, *op. cit.*, p. 134.

capsules. It is very pretty and sweet when in flower. The varieties of *Anabasis articulata*, whose bracts wither a showy red and rich claret colour, are common here. This species is quite abnormal at this season, having shed all its more slender twigs, and having more the habit of a Zygophyllum. It was not till I reached Wâdy 'Arabah that it occurred in its natural form.

Lepidopterous insects were more numerous in these cooler stations, chiefly attracted by the tent lights at night. Of the earlier desert plants, Reaumaria and Gymnocarpum are still abundant.

Several grasses, Cucurbitaceæ and Zygophyllaceæ belong to lower districts, but Fagonia ranges everywhere so far. *Ruta tuberculata*, with its disgusting smell, is still to be met with.

At 'Ain Suweirlyeh, where we camped for the ascent of Mount Sinai, there is a poor little garden containing pomegranates, palms, and nubk (Zizyphus), apricots, and mallow. Gomphocarpus is abundant about this well. It is one of the most remarkable species in Sinai.

I made the ascent of Jebel Mûsa and Jebel Katharina on November 20. On the way to the convent of Mount Sinai occurred *Centaurea scoparia*, Sieb.; *Celsia parviflora*, Dcne.; and *Alkanna orientalis*, Boiss. At the convent garden, where we dismissed our camels, are cypress, orange, figs, olives, dates, and vines in cultivation. These I only saw over the garden wall, for the delay in the convent was irksome since the whole thing was to be done in a day. On the garden gate were suspended several dead Egyptian vultures, which surprised me, as I thought the bird was too much valued as a scavenger to be destroyed. Gomphocarpus occurred again a little above the convent, which stands at 5,024 feet above sea-level. The following were first met with here: *Asperula sinaica*, Dcne.; *Pulicaria crispa*, Forsk.; *Verbascum sinaiticum*, Bth.; *Plantago arabica*, Boiss.; *Phlomis aurea*, Dcne.; *Nepeta septem-crenata*, Ehr.; *Mentha lavandulacea*, Boiss.; *Teucrium polium*, L., var. *sinaicum*; *Origanum maru*, Linn., β *sinaicum*; *Ficus pseudosycomorus*, Dcne.; and *Adiantum capillus-veneris*, Linn. A single tree stands near the spring, but I unfortunately lost my leaves of it It was, I believe, *Salix safsaf*, Forsk.

At this height, about 5,500 feet, a couple of palms (across the valley), *Phœnix dactylifera*, Linn., and a tall cypress, *Cupressus sempervirens*, Linn., var. *pyramidalis*, occur. The latter, which is not native, occurs a little higher, in a conspicuous place familiar to all travellers.

Cotyledon umbilicus, Linn.; *Arenaria graveolens*, Schreb.; *Scirpus holoschœnus*, Linn., *Peganum harmala*, Linn.; *Echinops glaberrimus*, D.C.; *Acanthodium spicatum*, Sieb., and several mosses were gathered on the ascent. On such occasions as these the Bedawîn made wild gestures and howls as I escaped from them into gullies and up cliffs. One reason of this I found to be their horror of boots, which they think most dangerous to the climber. At the second pyramid, that of Cephren, at Cairo, where I stole a march and reached the summit alone, the Bedawîn who pursued me made frantic efforts to deprive me of my boots ere the descent began. I need hardly say I valued the skin of my feet too highly to obey.

In spite of the Bedawîn, I followed the bent of my own botanical inclinations. The mosses were the result of a detour from the beaten track to a less open gully looking north. On or close to the summit, 7,320 feet, were *Cratægus sinaica*, Boiss.; *Artemisia herba-alba*, Asso.; *Verbascum sinaiticum*, Bth.; *Ruta tuberculata*, Forsk.; *Peganum harmala*, Linn.; *Arenaria graveolens*, Schreb.; *Buffonia multiceps*, Dcne.; *Poa sp.* (*P. sinaitica*?); and *Ephedra alte*, C. A. Mey., and others not recognisable. The ascent to the summit from the convent occupied about two hours.

The most striking feature in the aspect of the flora of the upper parts of Jebel Mûsa, from the convent upwards, is the prevalence of the Labiate and Scrophulariaceous families. Several fresh species had appeared, some of these peculiar to Sinai, and others seen before were very abundant here. As these orders increase, the Compositæ, abundant at intermediate heights, diminish towards the upper zone. The fern and the mosses illustrate the cooler atmosphere of the elevated region, though their immediate existence depends on unfailing springs of water. Having left our party here, I descended rapidly to the convent of Deir el 'Arbaîn, about 1,700 feet below, in the bottom of the gorge between Jebel Mûsa and Katharîna. With a nimble Arab as guide we did this in half-an-hour. At the convent I was transferred to another native. There was barely daylight left in which to accomplish Jebel Katharîna. I had arranged that my camel should be in readiness here to bring me back to camp at 'Ain Suweirîyeh at night. A quarter of an hour after my arrival the faithful Khalîl appeared, and I started at once—1.30 p.m.—for the summit.

At the monastery, or near it, were *Bupleurum linearifolium, D.C.,* var. *Schimperianum, Boiss., Carum sp.?; Pterocephalus sanctus, Dcne.; Veronica syriaca, J. et S.* (introduced); and *Celsia* and *Anarrhinum* already mentioned. *Salix safsaf, Forsk.,* occurs here. During the ascent most of the labiates and the hawthorn of Mount Sinai were met with; but this mountain wore a far more wintry aspect than its lower neighbour. A lack of running water renders it at all seasons more barren. At the spring Mayan esh Shunnâr, 'fountain of the partridge,' I made another little gathering of mosses, in all from the two mountains ten species, *i.e.*: *Grimmia apocarpa, Linn.; G. leucophæa, Grev.; Gymnostomum rupestre, Schwæg.; G. verticillatum; Tortula inermis, Mont.; Eucalypta vulgaris, Hedw.; Entosthodon templetoni, Schwæg.; Bryum turbinatum, Hedw.; Hypnum velutinum, Linn.; H. ruscifolium, Neck.* These are all British species, with the exception of *Tortula inermis,* which occurs also on the Morocco mountain at 8,000 to 10,000 feet, and no doubt elsewhere round the Mediterranean. One only in the list, *Gymnostomum rupestre,* is sub-alpine in Great Britain. There are two other mosses also, common British species, recorded from Mount Sinai by Decaisne.

The remainder of the ascent was over barren and perfectly unvegetated rock. Nevertheless, within a few hundred feet of the summit I was rewarded by finding the exquisite little *Colchicum Steveni, ? Kth.,* of a delicate pale lilac colour, sometimes white. It had no leaves, and bore either one, two, or three flowers on the scape; usually only one. It occurred again on the extreme summit, and I secured several bulbs. *Colchicum Steveni* was gathered afterwards on Mount Hor, where the flowers were very decidedly smaller. The Jebel Katharina plant may prove to be specifically distinct. This Colchicum has been recorded from the Palestine coast as far south as Joppa.

On the summit there was hardly any life. I obtained *Buffonia multiceps, Dcne.; Arenaria graveolens, Sch.; Herniaria, sp.? (H. hemistemon?); Gypsophila hirsuta, Led.;* and *G. alpina, Boiss.,* and fragments of an Astragal, perhaps *A. echinus, D.C.* On the ascent I gathered the root and leaves of a sedge looking like *C. distans, Linn.*

The summit of Jebel Katharina, 8,536 feet, the highest in the peninsula, was very cold, barely above freezing-point. Its mean annual temperature would perhaps about correspond with that of Edinburgh, while Jebel Mûsa would be nearer that of London. It is a solid hump of syenite with a lower shoulder joining it to a similar prominence about half a mile away. The view was magnificent, including the whole coast-line of Sinai from Suez to 'Akabah, except the portion intercepted by the Umm Shomer range to the south, whose summit almost equals that of Jebel Katharina. Jebel Mûsa looks a mere trifle, one of a fierce sea of red pointed and serrated peaks and ridges.

The summit was reached at 3.15, left at 4, and the convent of Deir el 'Arbain regained at 5. A long camel ride through a wild gorge by moon-light brought a memorable day to a close.

In the gorge I heard a deep clear strange note, which my Bedawin called 'hôadoô.' It seemed to proceed from an owl, and may have been Bubo ascalephus, the Egyptian eagle owl; but, much as my curiosity was aroused, there was no means of gratifying it.

With the exception of a couple of chats (*Saxicola leucopygia*, Br., and *S. lugens*, Licht.), and the Egyptian vulture, no birds were seen. A single coney (*Hyrax syriacus*, H. et Ehr.) showed himself for a few seconds on the summit of Jebel Mûsa.

CHAPTER IV.

MOUNT SINAI TO 'AKABAH.

OUR journeyings from Mount Sinai lay east of north to 'Akabah, skirting and occasionally crossing corners of the Tîh plateau.

Hares were occasionally seen of the little long-eared Sinaitic kind, and gazelle tracks were very numerous in Wâdy Zelegah (Zolakah). The lizards already mentioned are plentiful in this wâdy, and several geckos were captured, which proved to be of two species. A snake, *Zamenis ventrimaculatus*, was safely lodged in my spirit cylinder.

Wâdy Zelegah is a noble valley plain, about half a mile wide for upwards of twenty miles, bounded by precipitous cliffs and mountains. Several detours were made into the Tîh cliffs on the left of our line of march. The chief plants were—*Glaucium arabicum*, Fres.; *Capparis galeata, Fres.; Cleome arabica, Linn.; Ruta tuberculata, Forsk.; Odontospermum graveolens, S. Bip.; Artemisia herba-alba, Asso.*, and vars.; *Sonchus spinosus, Forsk.; Verbascum sinaiticum, Bth.*; and for the first time *Moricandia dumosa, Boiss.; Capparis spinosa, Linn.; Iphiona scabra, Del.*, and *Imperata cylindrica, Beauv.*

Frequent bags of fossils were obtained *in situ* for the assistance of the Geological Survey.

In birds, the white wagtail and the little cock-tailed wren-like warbler (Drymœca) were the most frequent. Desert larks and shrikes also occurred at scattered intervals. A very small warbler, *Sylvia nana*, was shot amongst tamarisk bushes. The song of the Drymœca is quite wren-like, but less piercing.

The flora is that of the western side: Tamarix, Caylusea, Retama, Ochradenus, Zilla, Santolina, Artemisia, Ærua, Ballota, Stachys, Lavandula, Anabasis, of species already mentioned, predominate. Several of the Mount Sinai groups of labiates were missing, as also two or three of the Iphiona group of composites. The larger Capparis was very frequent, growing on the most arid rocks above the wâdy flats, where nothing else, except perhaps *Lavandula coronopifolia, Poir.*, appears able to exist. *Capparis galeata* is sometimes an erect shrub 6 or 8 feet high, of a bright green, differing from the slender trailing blue-foliaged species, C. spinosa, which often grows with it. The former was now in fruit, the latter barren.

Camels delight in the larger grasses, in Ochradenus, Zilla, Nitraria, Anabasis, and tamarisks.

At the head of Wâdy el 'Ain, a grove of tamarisks was plentifully indued with an excrescence or exudation of grayish-white pilules of a viscid substance, with a faint taste of nucatine. This is the so-called 'manna of Sinai,' which is, I believe, more plentifully obtained from *Alhagi maurorum, D.C.* This gum is said to be due to the puncture of a small insect.

Life became more plentiful. Three butterflies were observed: a pale blue, a sulphur-yellow with brown under wings, and an admiral. Hornets and a long-bodied insect darted about in a broiling sun. I obtained all these except the sulphur-yellow butterfly.

In plants *Suæda monoica, Fres.*, and for the first time the rare *Linaria macilenta, Dcne.* This spring species was in flower, but the fugaceous corolla falls at the slightest touch. *Cleome droserifolia, Del.*, was also here first obtained. A spring supported a stream that moistened the soil for about a mile ere it gradually died a natural death. It led us the way into an unexpected and magnificent fissure in the red granite, the Wâdy el 'Ain. For five or six miles the gorge passes between sheer cliffs of this richly-coloured rock, with a height varying from 500 to 800 feet, and from 10 to 50 yards wide. It is in some ways the most impressive natural feature I have ever beheld. The floor is hard and level, and as the sun rarely hits the base of the cleft, many plants remained here in a fresher condition than elsewhere, and some new varieties were found. I will mention the less common species procured in this remarkable 'Sîk,' or cleft, which has rarely been visited : *Moricandia sinaica, Boiss.; M. dumosa, Boiss.; Cleome droserifolia, Del.; Capparis galeata, Fres.; Abutilon fruticosum, G. et P.; Zygophyllum coccineum, Linn.; Tephrosia purpurea, Pers.; Pulicaria (Francœuria) crispa, Forsk.; Blumea (Erigeron) Bovei, D.C.; Iphiona scabra, Del.; Sonchus (Microrhynchus) nudicaulis, Linn.; Scrophularia deserti, Del.; Linaria macilenta, Dcne.; Lycium arabicum, Schw.; Hyoscyamus aureus, Linn.; H. muticum, Linn.; Ballota Schimperiana, Bth.; Teucrium sinaicum, Boiss.; Origanum maru, Linn.,* β *sinaicum, Boiss.; Atriplex leucoclada, Boiss.; Typha angustata, B. et C.; Cyperus lævigatus, Linn., et var. junciformis; Panicum turgidum, Forsk.; Pennisetum dichotomum, Del.; Imperata cylindrica, Beauv.,* and forms of *Reseda pruinosa, Del.; Fagonia cretica, L.*, as well as other indeterminable remains. Several of the above are peculiar to Sinai, and some mentioned here and elsewhere are now first included in its flora.

It was with misgivings we camped in this wâdy. Had a 'seil' like the Rev. F. Holland's memorable one at Feirân visited us, we would have assuredly had a bad time. But the expected rain did not yet arrive.

While we were encamped here we received notice of the arrival of visitors for whom our ever-courteous chief prepared coffee. The party, consisting of engineers, Colonel Colvile, I believe, and others, passed us at speed on the opposite side of the narrow valley without a greeting. Suspecting that this impetuous haste, and absence of that courtesy for which Englishmen on their travels are so justly famous, arose from ulterior motives, Professor Hull summoned a council of war, which resulted in despatching our able conductor, Bernard Heilpern, with orders to secure our entitled priority to the 'Akabah Sheikh's camels and services. Bernard passed the fugitives in the night, and was entirely successful.

It was long ere we got clear of this ever widening, slowly rising Wâdy el 'Attîyeh, which wound through granite hills and lifted us out of Wâdy el 'Ain. Our height above sea-level varied between 2,500 and 3,000 feet. Acacias are numerous, chiefly *A. seyal*. This small tree, when not too flat-topped, as is commonly the case, has at a little distance a close resemblance to our hawthorn, with its gnarled and twisted stem and rugged bark. The granite hills, usually capped with a stratum of sandstone, are barren in the extreme. *Dæmia cordata* and *Tephrosia purpurea* are the only noteworthy species.

Hey's sand partridges were frequent, and good to eat. All seen as yet were of the one species. They rarely fly until almost walked on, trusting for escape to their close resemblance in colour to the shingle and rocks they inhabit. Until they run, which they do with rapidity, they would be most difficult to observe. Nevertheless they often betray themselves by their sharp cry of alarm. The Bedawîn then, swift, stealthy, and barefooted, gets easily amongst them, for they seem more alarmed by a noise than by the human figure. The Bedawîn flint lock is, however, slow and dignified in its performance, and usually affords abundant time for escape from its uncertain discharge.

Rock-pigeons and martins (*Columba Schimperi, Bp.*; *Cotyle rupestris, Scop.*) were seen in Wâdy el 'Ain.

All about the caper is frequent. The Arabs eat the ripe red fruit and seeds. I tasted it, but did not continue to eat it. The skin is like mustard, and the seeds like black pepper.

In a marshy place at the head of Wâdy el 'Ain, amongst palms and tamarisks, Typha angustata was 12 to 14 feet high; Erigeron Bovei 6 or 7 feet high, well branched and with many flowers, and Phragmites gigantea was fully 15 feet high.

The pricklier plants, Acacias, Acanthodium, Gymnocarpum, etc., are commoner, generally speaking, on the granite and sandstone than on the limestone.

In a very dirty well, Bîr es Saura, near the base of Jebel 'Arâdeh, *Chara hispida, Linn.*, occurred, and with it *Juncus maritimus, Lam.*, β *arabicus*.

On the summit of Jebel 'Arâdeh there was no vegetation, and in the limestone now lying above the sandstone numerous cretaceous fossils were obtained. A single white butterfly (Pieris sp.) was the only living thing. I estimated the height of this mountain 3,400 feet. It is about 1,300 feet above the plain, and forms a most conspicuous object. Like others, except those of granite, in this region, it is crumbling away and turning to dust on all sides. The beds of chalk and flints are much disintegrated, while all the outer surface of the lower limestone is on the move.

The only plants were Gymnocarpum, Reaumuria, Capparis, Acanthodium, Lavandula of the usual kinds.

We were here in a little known and unsurveyed region. Consequently there was abundant work for the engineering section of our party. Very few travellers had passed this way since Laborde's time, and I was sorely disappointed to find on the tableland we were now entering there was little living vegetation, although abundant withered evidence of a sparse but varied flora.

This tableland is called here Jebel Herteh, and is, properly speaking, a portion of the Tîh plateau, which becomes indefinite at its south-eastern border. A fine oval plain, Wâdy Hessi, about three to five miles broad, literally abounded in lizards, and here I killed another Zamenis, a sand-coloured snake about 4 feet long. A large-headed Arachnid (Sparacis sp.) was also very abundant, and seems to form food for some of the numerous chats and larks. Small flocks of sparrows, *Passer hispaniolensis, Temn.*, occurred here, while there was usually a raven or a vulture in sight.

This wâdy, now clad with withered scraps, is a favourite pasturing place later on for the Bedawîn flocks. I gathered here *Tribulus alatus*, D.C.; *Anastatica hierochuntina, Linn.*; *Zygophyllum dumosum, Boiss.*; *Lotus lanuginosus, Linn.*; *Ifloga spicata, Forsk*; *Filago prostrata, Parlat.*; *Linaria floribunda, Boiss.*; *Verbascum sinuatum, Linn.*; *Heliotropium undulatum, Vahl.*; *Micromeria myrtifolia, Boiss.*; *Plantago ovata, Forsk.*; *Panicum Teneriffæ, R. Br.*; and *Aristida cærulescens, Desf.* These had not been previously met with. Other interesting species not recently seen were *Farsetia ægyptiaca, Turr.*; *Reseda prunosa*,

Turr.; Polycarpæa prostrata, Dcne.; Helianthemum Lippii, Pers.; Atractylis flava, Desf.; Zygophyllum album, Linn., and others of commoner sorts.

In these depressions of the plateau, where water and soil are of more frequent occurrence, there is an abundance of grayish scrub, short, thin, and interrupted, and composed chiefly of *Zygophyllum dumosum, Anabasis (Salsola) articulata, Ephedra alte* and Atriplices, Nitraria, Zilla, Retem, and sometimes tamarisk.

Sonchus nudicaulis, Linn.; Dæmia cordata, Br.; Gomphocarpus and Lindenbergia still occur.

I endeavoured to obtain the Arabic names of the commoner species, and to confirm them from the mouths of two or more Bedawîn. These names so obtained rarely agree with those I find quoted in Forskahl, Boissier, Tristram, and others. It is probable that every tribe has its own plant-names.

An Arab informs me that 'boothum,' a tree growing on Jebel Serbâl and nowhere else, with a stony fruit, is used, its leaves being boiled as a cure for rheumatism, an infirmity to which the Arabs are martyrs. I suspect the plant to be *Cratægus aronia*. Also that safsaf (*Salix safsaf, Forsk.*, or *Populus euphratica, Linn.*) is the wood in demand for charcoal to colour their gunpowder. This they obtain in the valley between Jebel Mûsa and Jebel Katharîna as well as on the latter mountain. The proportions of their gunpowder are—one part sulphur, four parts saltpetre, and a little charcoal to colour.

Anastatica hierochuntina, Linn., 'Kaf Maryam,' or Rose of Jericho, was first seen here, and becomes common to 'Akabah and northwards to the Ghôr es Safieh. Ephedra alte is the most characteristic and abundant species. Acacias are almost absent. We were on a limestone tableland with occasional outcrops of sandstone. Once on such an outcrop a single shrub of *Acacia seyal* occurred. In exposed situations these acacia bushes, formed like a table, with its single leg much nearer one side than the middle, point with their overhanging part in the direction of the prevailing wind. On reaching the granite pass into 'Akabah the acacias again become abundant, but their absence above may be partly explained by the exposed situation.

Camels eat even the milky asclepiads, as Dæmia, which is said to be highly poisonous. *Heliotropium arbainense, Fres.*, was first met with by the Haj route from Cairo to 'Akabah, which we were now close to.

Those two especially nauseous species, Peganum and Ruta, are very frequent. The smell of the former is like that of our hound's tongue; the latter reminded me of some kind of wood-bug, which I collected in an evil moment in the scaffolding of the Milan Cathedral. *Cleome droserifolia, Del.*, smells like a fox. Other species here are *Malva rotundifolia, Linn., Linaria macilenta, Dene.; Deverra tortuosa, Gærtn.;* and *Ærua javanica, Juss.*

On November 29 we descended a magnificent gorge between granite and limestone by the Haj road to 'Akabah, which takes its name ('Akabah, 'steep descent') from this entrance. The ever-varying peeps of the gorgeously blue gulf of 'Akabah, shining in an intense sunlight, were a most refreshing change from the desert. The rich purple colouring of the lofty mountains of Midian formed a noble background.

CHAPTER V.

'AKABAH.

AT 'Akabah we remained from November 29 to December 8. I increased my collection here considerably. The flora displayed several fresh species. Bird life was more plentiful, and a large collection of shells was made on the beach. These, consisting of upwards of 200 species, including those from Suez, I have had determined by Mr. G. B. Sowerby, and amongst them are many which do not appear to have been admitted as inhabitants of the Red Sea.

'Akabah, even at this season, was oppressively hot. A swim in the sea, or rather a crawl amongst the coral reefs, about 3 feet below the surface, was delightful. Farther out sharks abound.

The straggling Arab village lies at the south-eastern corner of the plain which forms at once the head of the gulf and the southern end of the Wâdy 'Arabah. This is the narrowest part of the wâdy, being not more than five or six miles across.

A very fine tree of *Acacia tortilis, Hayne.*, stands close by. On the coast are many clumps of the date palm, interspersed with a very few trees of the doum palm (*Hyphæne thebaica, Del.*), already noticed here by Mr. Redhead. The doum palm, a native of tropical Africa, Nubia, and Abyssinia, finds its northern limit at 'Akabah.

In the enclosures here I noticed nubk (Zizyphus), henna (Lawsonia), palms, tamarinds (*Tamarindus indica, L.*), pudding pipe (*Cassia fistula ?*), figs, and several kinds of gourds. Most esculents were still invisible or in a seedling state.

There is but one boat at 'Akabah. Laurence and I succeeded in hiring this with a native fisherman, with two Arabs, nets and lines. There were many flying fish (Exocœtus) about. We first rowed across the corner of the gulf to the sandy beach, where the two Arabs landed, and with a circular casting net captured some small fish ('Akadi' and 'Sahadan') for bait. With these and some loose stones, about a pound weight each, we rowed out a few miles. The bait fish, broken in three, is affixed to the hook and one of these stones is hitched to the line a little above with a slip-knot. On reaching the bottom a couple of violent jerks dismiss the sinker and let the line swing free. We caught fish rapidly, 'hedjib,' at Suez called 'jar,' 'gamar' (a species of Chætodon ?), and one splendid red fish they called 'bossiah,' without scales, and very good to eat. We also hooked a shark, 'Zitani,' about 5 feet long, who amused us for a time and then carried off the line.

Before dismissing our Towarah Bedawin I had endeavoured to pump them of what little information they possessed about the feral inhabitants of Sinai. They knew of leopards on Serbâl and Umm Shomer; wolves in Wâdy Lebweh and neighbourhood; hyænas, ibexes, gazelles, hares, jerboas, rats, and mice made up their total. Their sheep they say were imported from Arabia; they have a few donkeys and camels; their goats are a distinct breed, which they are especially proud of. Five kinds of snakes they admitted, all of which were poisonous! The one I caught in Wâdy Zelegah, *Zamenis ventrimaculatus*, attains a full size of 5 or 6 feet. These remarks I set down to be taken for what they are worth.

Dr. Hull captured a handsome little snake here, and handed it over to me; it proved to be *Zamenis elegantissimus*, and is now in the British Museum.

The birds obtained at 'Akabah were: *Cercomela melanura, Temn.;* *Cyanecula cærulescens, Pall.;* *Argya squamiceps, Rupp.;* *Motacilla alba, Linn.;* *M. flava, Linn.;* *Pycnonotus xanthopygus, Hemp. et Ehr.;* *Lanius fallax, Finsch.;* *Passer hispaniolensis, Temn.;* *Ægialitis asiatica, Pall.;* *Tringoides hypoleucos, Linn.,* and several larks and chats already mentioned. Ravens, crows, martins, rock-pigeons and the little gull, *Larus minutus, L.*, were also observed. Vultures and English swallows were frequently to be seen, the former usually of the Egyptian species.

Not many identifiable plants occurred here which had not been previously seen. These are: *Cassia acutifolia, D.C.; C. obovata, Coll.; Onobrychis Ptolemaica, Del.; Tephrosia apollinea, Del.; Artemisia monosperma, Del.; Statice pruinosa, Linn.; Salvia deserti, Dcne.; Boerhavia plumbaginea, Cav.; Calligonum comosum, L'Her.; Atriplex crystallina, Ehr.,* and *Andropogon foveolatus, Del.* A few other less common species may also be mentioned : *Lotononis Leobordea, Linn.; Tephrosia purpurea, Pers.; Sonchus spinosus, D.C.; Cucumis prophetarum, Linn.; Linaria macilenta, Dcne.; Trichodesma africanum, R. Br.; Heliotropium arbainense, Fres.* Forskahlea, Andrachne, Panicum, and others. Along the shore in some places is a close growth of *Nitrara tridentata, Atriplex leucoclada, Boiss.; A. halimus, Linn.; Juncus maritimus, Linn.;* var. *arabica,* and others. *Cressa cretica* is a characteristic species along the shore on the saline flats.

Gathering shells where such an abundance of, to me, novel forms occurred, was enthusiastically pursued. I shall not here deal with this subject in any detail, but merely mention the principal genera met with. These were mostly univalves, bivalves being scarcer in species, and infinitely fewer in individuals. Great numbers of opercula of a Turbo, pretty polished little hemispherical bodies retaining the spiral lines of structure, pens of calamaries, and the delicate vitreous wingshells of pteropods occurred, as well as a large variety of fragments of coral. Conus, Cerithium, Strombus, Cypræa, Mitra, Triton amongst univalves; Arca, Pectunculus, Tridacna, Chama, and Venus amongst bivalves, were the best represented genera. Drift shells are rarely disturbed, the tide being apparently not above a foot in range at 'Akabah.

CHAPTER VI.

'AKABAH TO MOUNT HOR.

At 'Akabah we have left the Sinaitic peninsula; from here we turned northwards up the Wâdy 'Arabah. Happily we had occasion henceforth to travel more slowly, in order to give the surveying party time to keep pace with us. I was thus enabled to make wide detours east and west out of the 'Arabah, but my inclination lay chiefly eastwards into the precipitous borderland of Edom.

In the Wâdy 'Arabah I saw gazelles several times; Wâdy Menaiyeh, on the west, may be mentioned as a good hunting-ground. These graceful animals seemed more at home on the west side, abounding on the Judæan wilderness, and all over the Tîh plateau. Ibexes, on the other hand, appeared more frequently on the higher mountain declivities of Edom, to the east. Hyænas, judging from their tracks, must be plentiful; once I had a good view of one, and quickened his lolloping pace with a fusilade from revolver and fowling-piece. At El Tâbâ, on the east side, about twenty-five miles north of 'Akabah, a fruitful, marshy place with a deep spring, I saw perfectly fresh tracks of 'nimr,' or leopard, and subsequently, at 'Ayûn Buweirideh, Laurence came on fresh remains of some beast which had served apparently as a meal for these animals. A hare, the Sinaitic species, was killed a few miles north of 'Akabah. A much larger hare, *L. ægyptiacus*, was seen several times on the eastern declivity of the Tîh. My frequent failure in bringing down game and specimens I attributed partly to my having been unable to land English cartridges or powder in Egypt, and being dependent on very worthless and very expensive ones procured in Cairo. I would recommend all sporting travellers to run any risk in smuggling sooner than let this occur to them.

The Wâdy 'Arabah abounds in rodents. These animals appear to be chiefly nocturnal in their habits, and are very seldom seen. The number of holes and the abundance of their tracks is truly astonishing. Their colours are usually in strict harmony with the desert, for the Wâdy 'Arabah is some ten to thirteen miles across, and more correctly called a desert than most parts of Sinai. Jerboas were seen a few times, and Gerbilles, of which I trapped one, appear to be most numerous.

Birds increased in numbers and variety. From El Tâbâ northwards, about twenty-five miles from 'Akabah, a grove of acacias (chiefly *A. tortilis, Hayne.*), and a little Zizyphus, stretches about ten miles along the eastern edge of the 'Arabah. A smaller grove occurs nearer 'Akabah, at the mouth of Wâdy el Ithm, where I first met with the 'hopping-thrush.' In the larger grove the handsome *Loranthus acaciæ, Zucc.*, abounds.

Several times I endeavoured to get a shot at a small bird here which uttered a sharp little note, new to me, but I was unsuccessful. Mr. Armstrong, who was with me that day, and is well skilled in Palestine birds, recognised it, having also seen the bird, as the little Sunbird, *Cinnyris Oseæ*. Subsequently, when I reached the Ghôr, I obtained several specimens and recognised the note at once. This species has not been detected south of the Ghôr, where it was first made known, like the hopping-thrush, by Canon Tristram.

The Sunbird probably follows the Loranthus, to whose flowers it appears attached. Its long bill reaches the base of the tubular flower, searching for honey, and it thus probably secures their cross-fertilization. One was shot in the Ghôr in the act of doing so, its bill being covered with the pollen of the Loranthus.*

The hopping-thrush (*Argya Squamiceps*) is a remarkably weak flier, hardly leaving the ground except in tremendous jumps, which cause his large fan-shaped tail to overbalance and almost overturn him as he makes a pause. He is a most grotesque bird; nevertheless the mournful cries of one when I had shot his mate impressed me with a different feeling.

Palestine bulbuls were occasionally seen here also. Hooded chats, Persian larks, and desert larks were frequent, and large flocks of sparrows assembled about us in several places.

The floor of the wâdy is sometimes alive with geckos, lizards, and ants, as well as numbers of long-winged males of a Persian species of white ants, *Hoeotermes vagans, Hag.*, not yet able to fly, over which the hopping-thrushes fall into inconceivable excitement.

* Since writing the above I find that Burton has seen the Sunbird, almost certainly this species, about five degrees from this southwards, in Midian. 'Land of Midian,' vol. ii.

The first bee I met with was captured here, and small beetles are often sacrificed to the good of science. I spare the reader the enumeration of their scientific names, which will be given fully at the close.

At El Tâbâ occurred a greensward of *Cynodon dactylon, Linn.* In or near the grove already spoken of were *Cocculus Leæba, D.C.; Fagonia myriacantha, Boiss.; Scrophularia deserti, Del.; Loranthus acaciæ, Zucc.; Salsola fœtida, Forsk.; Eragrostis cynosuroides, Retz.,* and common sorts. In the open sandier wâdy, *Glaucium arabicum, Fres.; Gypsophila Rokejeka, Del.; Monsonia nivea, Dcne.; Microrhynchus nudicaulis, Linn.; Iphiona scabra, Del.; Citrullus colocynthis, Schr.; Cleome droserifolia, Del.;* Cucumis, Pancratium, Danthonia, Trichodesma, Andrachne, Forskahlea, Anabasis, and Tamarisk form almost the whole vegetation.

In some places the wâdy is spanned by rolling wastes of sand dunes 10 to 12 feet high. These appear to have been formed around the bases of clumps of tamarisk and anabasis, which is here very tall, 6 to 8 feet high or more.

Ochradenus baccatus is very abundant, often overtopping the acacias by whose protection from camels it thrives. Lycium europæum and one or two grasses escape being cropped in the same manner, and grow to an unwonted size.

On December 7 a long day's climbing with Laurence brought us to the head of Wâdy Ghurundel in Edom. This was at a height of about 1,800 feet above sea-level, six miles east from the 'Arabah. The scenery on the way was superb. Huge blocks of red sandstone, 800 to 1,000 feet high, towered above us, sometimes sheer and tottering in broken masses from the main cliffs behind. We passed a spring with a few date-palms, and a little higher a large bulb with broad leaves (*Urginea scilla, Steinh.?*) first appeared and soon became abundant. It was not yet in flower. *Dianthus multipunctatus, Ser.; Eryngium, sp.; Odontospermum pygmæus, Cav.; Cotula cinerea, Del.; Solanum nigrum, Linn.* (var. *moschatum); Satureia cuneifolia, Ten., forma; Boerhavea verticillata, Desf.; Ficus sycamorus, Linn.; Traganum nudatum, Del.; Aristida ciliata, Desf.,* appeared for the first time. The Odontospermum (*Astericus*), which occurred at a considerable height, was a little woody button representing the hardened flower head, which was usually solitary and close to the ground. This plant, like Anastatica, has hygrometric properties, and has been put forward by Michon as the true Rose of Jericho of the travellers of the Middle Ages. *Anastatica hierochuntina* will not, however, be readily deprived of its claims.

Besides the above, which were all gathered farther on, some plants of more limited range occurred: *Moricandia dumosa*, Boiss.; *Abutilon fruticosum*, G. et P.; *Varthamia montana*, Vahl.; *Iphiona scabra*, Del.; *Centaurea scoparia*, Sieb.; *Iphiona juniperifolia*, Coss.; *Ballota undulata*, Fres., and others already met with.

Judging from the abundance of its bur-like carpels lying in the dry watercourses, *Calligonum comosum* is the most abundant shrub; it was now in a withered condition. Several other bulbous species which occurred here are as yet undetermined. A stiff scramble brought me back to the 'Arabah by a more northern valley. Amongst land shells, Helices of four species were gathered in Wâdy Ghurundel.

CHAPTER VII.

PETRA AND MOUNT HOR; WÂDIES HARÛN (ABU KOSHEIBEH) AND MÛSA; JEBEL ABU KOSHEIBEH.

THE last valley has shown us some characteristic Sinaitic species extending their range north-eastwards across the great valley of the 'Arabah. Several more will appear in the group of localities now to be considered. Were I to hazard a suggestion here, it would be that these plants, formerly considered peculiar to Sinai, have had their origin more eastwards, and have spread, like many other Arabian plants, in a westerly direction.

Owing to the greater moisture found in the upper part of some of the valleys of the Edomitic escarpment, there is a greater variety of species and a sprinkling of ferns, mosses, and lichens. These are mostly more northern forms, spreading southwards at high levels.

We are now entering a district which Canon Tristram has somewhat liberally included in Palestine. The flora has its own peculiar plants as well as a large proportion of southern or Sinaitic species, and thus it adds many to the Palestine flora. I will first speak of the wâdies, and then of Mount Hor and Petra. The latter places, I think, have not been botanized previously to my visit, and are visited only with difficulty and expense, owing to the cupidity and lawlessness of the sturdy beggars or Bedawîn who dwell there.

Irby and Mangles, Commanders in the Royal Navy, travelling in 1816-1820, were the first Europeans who visited these regions in modern times. Further on I will quote a few remarks from their most interesting volume, since I find no other allusions to the vegetation of the ancient capital of the Nabathæans.

The following plants not previously seen were gathered in Wâdy Abu Kosheibeh (Wâdy Harûn), and on the Jebel or peaked mountain which stands in a commanding position across its head: *Fumaria micrantha, Laq.; Erodium hirtum, Forsk.; Poterium verrucosum* (?), *Ehr.; Anvillæa Garcini, D.C.; Carthamus glauca, M.B.; C. lanatus, Linn.; Globularia arabica, J. et S.; Podonosma syriaca, Lab.; Nerium oleander, Linn.; Pentatropus spiralis, Forsk.; Boucerosia, sp. nov.; Salvia ægyptiaca, Linn.; Juniperus phœnicea, Linn.; Bellevallia flexuosa, Boiss.; Asparagus aphyllus, Linn.; Asphodelus ramosus, Linn.; Pennisetum cenchroides, Rich.; Cheilanthes odora, Sw.,* and *Notholæna lanuginosa, Desf.* Of these, Globularia, Podonosma, Boucerosia, Juniperus, and the two ferns were obtained above the wâdy amongst the cliffs of Jebel Abu Kosheibeh, from about 3,000 to 3,500 feet above sea-level.

The Globularia is a pretty compact little shrub, with blue heads of flowers and small entire leaves; the species here is the Arabian form, *G. arabica,* perhaps hardly distinct from *G. alypum, L.,* of the Mediterranean.

The two Asclepiads, Boucerosia and Pentatropus, are both frequent; the latter is probably *P. spiralis,* but as it was not in flower, Mr. Oliver would not speak positively. It occurred again at the Ghôr, trailing over acacias.

The Boucerosia may be *B. aucheriana, Dcne.,* an insufficiently described plant from Muscat in South-east Arabia, which is also the nearest known habitat for the Pentatropus.

On Jebel Abu Kosheibeh were also gathered: *Moricandia dumosa, Boiss.; Gomphocarpus sinaiticus, Boiss.; Helianthemum Lippii, Pers.;*

Cotyledon umbilicus (?), Linn.; *Linaria macilenta*, Dcne.; *Verbascum sinuatum*, Linn.; *Phlomis aurea*, Dcne., and *Boerhavea verticillata*, Poir. Many desert species of Reaumuria, Ochradenus, Zygophyllum, Morettia, Zilla, Acacia, Retama, Ruta, Ifloga, Lycium, Trichodesma, Forskahlea, Asphodelus, Anabasis, Ephedra, and grasses already mentioned, occur also in Wâdy Harûn, the name which the Bedawîn invariably give this wâdy.

It will thus be seen that there is no appreciable break as yet in the continuity of the Sinaitic flora as we travel up the Wâdy 'Arabah, but an increase of species from eastwards and northwards.

The Wâdy Harûn is at first wide and arid, but after a few miles vegetation rapidly increases with moister conditions. The flanks of the Edomitic limestone plateau are better supplied with moisture than the Sinaitic granite. Banks by the edge of this valley at a moderate elevation, 1,000 to 1,500 feet above sea-level, had a sparse coating of mosses and other cryptogams. The mosses were chiefly of the Tortula genus, of which five species were collected. Side by side with these grow the desert species above mentioned in great luxuriance. Dæmia cordata, for instance, climbed to a height of 10 or 12 feet in retem bushes; the support being as well developed as the climbing plant. In the open desert, Dæmia, as mentioned by Mr. Redhead, lies sprawling on the ground, its several stems sometimes closely twisted into a thong towards their extremity, so that all circulation is stopped, and the young shoots are strangled. This is probably due to changed conditions having deprived it of its normal support, which it rarely finds in the desert, and even seems there to have lost the power of utilizing. For I have seen it strangling itself side by side with bushes of the very sort which here gave it so much assistance. The desert plant was more plentifully milky, and we have here seen at work agencies which are giving rise to a modified form, in better harmony with its environment.

From the summit of Jebel Abu Kosheibeh, which I climbed with Dr. Hull, an unusual sight was observed: a stream, small in size, but containing a good body of water, rushing down the cliffs about half a mile to the south-eastward. I could distinguish with my spy-glass the growth of arundos and oleanders that fringed its banks, but unfortunately there was no time to examine it more closely. Running water was once seen before on Jebel Mûsa.

The juniper is a well-shaped bush or small tree, with a trunk sometimes a foot in diameter. It gives a considerable area of shade with its dark close foliage. A large specimen occurs immediately below the summit, and I could see it on all the highlands around, even at the summit of Mount Hor, which looked but a little distance off.

On December 10 we made the ascent of Mount Hor, returning to camp the same day by Petra. Our camp was fixed near the mouth of Wâdy Harûn. Although having made an early start (4 a.m.), the visit was necessarily a very hurried one. While waiting for a cloud to lift from the summit of Mount Hor for the benefit of the theodolite party, I had time, however, to make a good gathering of the bulbous plants, now just showing their leaves, with which the upper part of this mountain abounds.

The view from Mount Hor, whose height I estimated by aneroid at 4,400 feet, is a disappointing one, and bears no sort of comparison with those from the Sinai peaks. This defect is due to the adjoining high and monotonous tableland of Edom, which obscures one side of the horizon. This tableland averages perhaps 5,000 feet in height in the eastern neighbourhood of Mount Hor, and is composed of the unvarying and unpicturesque white cretaceous limestone. It lowers northwards, and I afterwards reached its outer edge. In some places it has quite a forest of vegetation.

With regard to Mount Hor, Irby and Mangles write: 'Much juniper grows on the mountain, almost to the very summit, and many flowering plants, which we had not observed elsewhere; most of them are thorny and some are very beautiful.'

As Mount Sinai is a mountain of labiates, so Mount Hor is a mountain of bulbs. The number of species and individuals of these orders respectively vividly coloured my impression of the botanical features of each of these sacred peaks. At the same time many of the Mount Sinai plants, labiates included, occur on Mount Hor. On Mount Sinai I procured bulbs of a single species, a total of three perhaps occurring. On Mount Hor I gathered at least twenty sorts.

In the upper 1,000 feet of Mount Hor a considerable accession of Mediterranean or more northern forms appear. A more interesting group is that of plants which have been considered absolutely peculiar to Sinai.

Both these lists, which I here append, would no doubt be swelled by observations at a more seasonable visit.

Northern species ranging south to Mount Hor:

    Dianthus multipunctatus, Ser.
    ? Geranium tuberosum, Linn.
    Pistacia palæstina, Boiss.
    Rhamnus punctata, Boiss., var. barren (sp. nov. ?).
    Paronychia argentea, Lam.
    Bryonia syriaca, Boiss.
    Galium canum, Reg.
    Scrophularia heterophylla, Willd.
    Sternbergia macrantha, Gay.
    Colchicum montanum, Linn.
    C. Steveni, Kunth. (also on Mount Sinai).
    Urginea scilla, Sternh.
    Bellevallia flexuosa, Boiss.
    Asphodelus fistulosa, Linn.
    Asparagus aphyllus, Linn.
    A. acutifolius, Linn.
    Arum, sp. (?).
    Carex stenophylla, Vahl.

No doubt many of these occur on the Edomitic plateau, whose botany is practically unknown.

Sinaitic species discovered on Mount Hor:

    Moricandia dumosa, Boiss.
    Pterocephalus sanctus, Dcne.
    Echinops glaberrimus, D.C.
    Varthamia montana, Vahl.
    Celsia parviflora, Dcne.
    Origanum maru, Linn., β sinaicum.
    Phlomis aurea, Dcne.
    Teucrium sinaicum, Boiss.

These have been considered peculiar to Sinai. They may now be included in the flora of Palestine.

A consideration of the latter group is especially interesting when considering the ancestral origin of the more local or endemic portion of the Sinai flora ; and it also gives us a slight clue to the probable nature of the flora of the little known region east and south-east of Mount Hor. Judging from an appendix of species of plants collected by Burton's expedition to 'The Land of Midian,' the flora of the upper regions of Sinai is more nearly allied to that of Edom to the north of east, than to that of Midian in the south-east. The Gulf of 'Akabah has formed a barrier in the latter case.

Of the bulbous species, here as elsewhere, I can only enumerate a portion. The bulk of those gathered were in leaf, and were brought home to Mr. Burbidge, of the College Botanic Gardens in Dublin, under whose care many retained life, but have not flowered.

The arboreal vegetation of Mount Hor was confined to the summit, and consisted of a bladder-senna, *Colutea aleppica, Lam.*; a turpentine tree, *Pistacia palæstina, Boiss.*, and a juniper, *Juniperus phœnicea, Linn.* Each of these was about 10 or 12 feet high. The Rhamnus already mentioned was very much stunted.

At Petra two new species were discovered, which will be described subsequently. One was a Galium allied to *G. jungermannioides, Boiss.*, and pronounced new by Mons. Boissier. It is a low straggling matted species, with the habit of our Asperula cynanchica. It occurred in the 'Sîk.' The other new species was a Daphne, an erect shrub 6 or 7 feet high, with long linear leaves, reddish-brown berries, and small cream-coloured flowers. The fibre is remarkably stringy and tough. The Daphne is allied to *D. acuminata* and *D. mucronata*, but differs materially from both these species. It occurred, in flower and fruit, on the slopes of Mount Hor, about a mile from Petra, and again at intervals lower down. The Boucerosia, already mentioned as being an undescribed species, was found on Mount Hor in flower in several places.

Many unrecognised fragments of Umbellifers, Scrophulariaceous plants, grasses, and others were noticed at Petra, and the botany will yield a good harvest to anyone arriving at a proper season, and with sufficient leisure. My time in Petra was somewhat under an hour!

The following plants, not previously met with, were gathered at Petra and Mount Hor: *Diplotaxis pendula, D.C.; Ononis vaginalis, Vahl.;*

Rubia peregrina, Linn.; Inula viscosa, Desf.; Zollikoferia casinianæ, Jaub.; Thymelæa hirsuta, Linn.; Salsola rigida, Pall.; S. inermis, Forsk.; Noæa spinosissima, Moq.; Polygonum equisetiforme, J. et S.; Allium sinaiticum, Boiss.; Asplenium ceterach, Linn.; Andropogon hirtus, Linn., in addition to those already mentioned as reaching here a southern limit, and the Abu Kosheibeh plants, which also, as a rule, occur on Mount Hor.

The majority of these additions occurred from about 3,000 feet to the summit. I extract a few notes from my journal on this subject.

At 3,000 feet Oleander and tamarisk cease, Scilla abundant; at 3,450 feet Thymelæa (Passerina) first occurs ; at 3,750 feet numerous species occur, as Pterocephalus, Globularia, Onosma, Juniperus, Ceterach, Cheilanthes, Fagonia, Cotyledon, Capparis spinosa, Varthamia montana, Phlomis, Ononis, Deverra, Moricandia dumosa, Rhamnus as I ascend; at or near the summit (4,400 feet about) are Geranium, Colutea, Pistacia, Pennisetum cenchroides, Hyoscyamus aureus, Noæa, Poterium spinosum, Scilla, Malva, Carex, Ephedra, Zollikoferia, Echinops, Verbascum sinuatum, Origanum, Ajuga tridactylites, Arum, sp., Bryonia, Sternbergia, and Colchicum, of species already mentioned.

Of Wâdy Mûsa, in which Petra is situated, Irby and Mangles write: 'Following this defile farther down, the river reappears, flowing with considerable rapidity. Though the water is plentiful, it is with difficulty that its course can be followed from the luxuriance of the shrubs that surround it obstructing every track. Besides the oleander, which is common to all the watercourses in the country, one may recognise, among the plants which choke this valley, some which are probably the descendants of those that adorned the gardens and supplied the market of the capital of Arabia : the carob, fig, mulberry, vine and pomegranate line the river side ; *a very beautiful species of aloe* also grows in this valley, *bearing a flower of an orange hue shaded to scarlet ;* in some instances it had upwards of one hundred blossoms in a bunch.' Several of these were not observed by us. Of the aloe I can give no information.

At Petra, 2,900 feet above sea-level by my aneroid, many of these and others occurred ; the most prominent were Phlomis, Ononis, Thymelæa, Rubia, Rhamnus, Pistacia, Inula, Sternbergia, Bellevallia, Rumex roseus,

Verbascum sinaiticum, Ficus sycamorus, and a stunted pinnate-leaved shrub or small tree, perhaps a Fraxinus. The Ononis, very viscid, with pretty yellow and claret-coloured veined flowers, was very abundant. So also was Thymelæa. Sternbergia (Colchicum) macrantha was glorious, with flowers of golden yellow as large as a lemon.

Few observations on animal life were obtained in this hurried visit, but these were all of interest.

Ibexes and gazelles were seen on Mount Hor, and a hare of the Egyptian variety fled from Wâdy Harûn at our approach. Another, seen at Petra, much lighter in colour, may have been the Nubian form.

When climbing Jebel Abu Kosheibeh, a clear loud flute-like whistle attracted my attention. The first few times I heard it I was fully persuaded it was a signal to warn those rascally Petra Bedawîn that hated Christians were invading their domain. But I presently saw the whistle belonged to a bird, which proved to be Tristram's Grakle. This species, originally discovered by Tristram about the Dead Sea, has since been found in Sinai at Wâdy Feirân by Wyatt, who also met it at Petra. All the time we were on this mountain several of these birds kept flying around us, often displaying the orange spot on the wing as they hovered close by. Their flight is very graceful, sometimes hovering butterfly-like, sometimes swift and undulating in large curves like the chough. Grakles were seen afterwards a little above Petra, and a flock of a dozen or thereabouts circled round the summit of Mount Hor, disappearing and reappearing from the corners of the red sandstone cliffs, and giving notice of their presence with their melodious whistle. This is probably a favourite breeding-place with these birds. It was not until I reached the Dead Sea that I obtained a specimen.

At Petra also occurred the Palestine bulbul, and the rich musical cry of the fantail raven, *Corvus affinis, Rupp.*, was almost incessant while we were there. Nevertheless this bird hardly came nearer than two or three hundred yards, and would be difficult to obtain. By its note and by its size, and by its broad expanded tail seen on the wing, I was assured of the species on referring to Canon Tristram's work. This raven and the grakle are two of that author's characteristic birds of the Dead Sea basin.

Hey's sand-partridge, shrikes, and desert larks are also not unfrequent, the latter lower down towards the 'Arabah.

To Laurence's sharp sight I was indebted for two snakes, *Zamenis Cliffordii, Schleg.*, and *Rhyncocalamus melanocephalus, Gunt.* The latter species was believed peculiar to the Jordan Valley, where it was found by Tristram, and forms as yet the single representative of the genus founded for it by Dr. Gunther. The former has not hitherto been found outside the African continent.

A centipede (Scolopendra) and a black millipede (Spirostreptus) four or five inches long, but fortunately torpid, were captured here. The latter seemed to be very common.

Wells, which I often searched with a net, yielded, as a rule, no life except small leeches and the larvæ of gnats. Some handsome insects of the grasshopper and cricket sorts were captured from time to time.

Up to this very few mollusca have been collected; *Helix seetzeni, Koch.*, and *H. candidissima, Drap* , were found in one or two places in Sinai. The latter was again met with in Wâdy Ghurundel in Edom, where I found also *H. prophetarum, Bourg.; H. filia, Mouss.*, and the handsome species *H. spiriplana, Oliv.* On Mount Hor this last was frequent, and another fine shell, *Bulimus carneus, Pfr.*, was here first found. Most of these became commoner down to the Ghôr. At Petra, and in the 'Arabah, I collected also *Helix cæspitum, Drap.*, a rare species. This scarcity of land shells is paralleled on the eastern side of the Gulf of 'Akabah in the land of Midian, where Captain Burton speaks of them as very rare, and mentions that he only met with two species in four months. In its natural history this little known country appears to be (judging from Captain Burton's work) almost identical with Sinai.

CHAPTER VIII.

WÀDY HARÛN TO THE DEAD SEA.

THE mouth of Wâdy Harûn into the 'Arabah is somewhat more than half-way from 'Akabah to the Dead Sea. The watershed between the Dead Sea and the Gulf of 'Akabah is nearer to 'Akabah. We estimated its lowest point at 660 feet above sea-level. It lies on the west side of the 'Arabah. At the mouth of Wâdy Harûn the 'Arabah is at its widest, being about thirteen miles across. The total distance from 'Akabah to the Dead Sea is 112 miles.

My chief detour in this part of the 'Arabah was on the east side, up a long valley to the Edomitic plateau, with Mr. Armstrong. On this occasion we returned to the 'Arabah by a more northern valley, Wâdy Ghuweir, which, from the numerous remains of encampments, tribe-marks ('Wasm'), and the well-worn tracks, appeared to be a leading thoroughfare into the Shobek country.

In this wâdy are several springs, appearing, as is frequently the case, at the union of the sandstone and limestone formations. One of these springs supported a jungle of reeds, with palms and some interesting composite species of luxuriant growth. Tamarisks, acacias, and nubk trees (Zizyphus) were in some profusion, and on each of these three trees the parasite, *Loranthus acaciæ, Zucc.*, with its handsome red flowers, was a conspicuous ornament. It was seen only two or three times on the tamarisk, oftener on the nubk, but much more usually on the acacia. Clinging to the reeds was an Asclepiad, *Cynanchum acutum, Linn.*, whose range is more Mediterranean than the others met with. Amongst them was the stately *Saccharum ægyptiacum, W.*, and a shrubby composite, *Pluchea dioscoridis, D.C.*, reached a height of 15 feet. Its flowers were insignificant. A red-barked osier, *Salix acmophylla, Boiss.*, and a poplar, *Populus euphratica, Linn.*, which is perhaps the willow of Babylon, occurred along the margin of the short-lived stream. Other species collected were : *Erucaria aleppica, Linn.; Tribulus terrestris, Linn.; Ficus carica, Linn.; Salsola tetragona, Del.*, and others less noteworthy. A very fragrant savory, *Satureia cuneifolia, Ten.*, and our early acquaintance the 'sekkaran,' *Hyoscyamus muticus, Linn.*, occurred.

At the head of this valley *Juniperus phœnicea* was found to be the tree visible from the 'Arabah on the white chalky plateau of Edom, and growing abundantly. Burton found this tree luxuriant and abundant at considerable heights in Midian three degrees farther south.

In this wâdy I gathered maiden-hair fern, the first I had seen since leaving Jebel Mûsa. Caper (*Capparis spinosa*), *Lycium arabicum*, and *Boerhavia verticillata* also occurred. Bushes of nubk were sometimes canopied with this latter trailing plant, with its pretty panicles of small blueish flowers.

The Bedawîn told me that with the juniper trees on Edom occur also 'balût,' *Quercus coccifera, Linn.*, and 'arour,' a thorn with a small sweet fruit. This was, I believe, *Rhus oxyacanthoides, Linn.*, which the above-mentioned traveller found abundantly in Midian. I met it subsequently in the Ghôr.

In Wâdy Ghuweir I captured the first Batrachian I met with, *Bufo viridis, Linn.;* running water, the rarest and pleasantest of sights in these regions, was the source of this increased variety of life.

At the 'Arabah, abreast of the above valley, I examined some large bushes of *Calligonum comosum, L. Her.*, a desolate, leafless, whitened, scrubby species which often grows in shifting sand. Its roots are beautifully adapted to secure its position. These are woody, springy, and tough, very different from the brittle branches, and about a quarter of an inch in diameter. Some of these are seven or eight yards in length, perhaps much more, and beset with knobs at intervals, which are serviceable in giving them a better grip. These excrescences may have been due to insects, for I afterwards noticed that this plant was much subject to galls; but whatever their origin, they served the purpose of the flukes of an anchor to hold the bush in a sea of shifting sand.

There appears to be a great variety of gall-producing insects in the desert. Almost every woody species is liable to knobs and swellings. One of the most curious of these appendages was that frequently attached to the common Salsola—a shapely little spurred and coloured excrescence like a solidified flower of one of our commoner wild orchids.

A minute cruciferous annual, half an inch high, leafless and with a silicle which formed almost the entire plant, was so fragile that it failed to reach home. The silicle valves had separated, dehiscing from the base upwards, one at either side of the septum.

In this part of the 'Arabah *Pancratium Sickembergeri* was frequently gathered. At the spring of 'Ayûn Buweirideh, a little south of Wâdy Ghuweir, I obtained many old friends. *Populus euphratica* attains here good dimensions. No less than three running streams maintain a brief but productive existence across the sands. I gathered here *Prosopis stephaniana, Willd.; Pulicaria arabica, D.C.; Statice pruinosa, Linn.; Artemisia monosperma, Del.; Suæda asphaltica, Boiss.; Salsola fetida, Forsk.*, and many more.

Several bulbous species were obtained here. One of these, which has flowered since my return, has been determined by Mr. Baker, *Urginea undulata, Desf.*

Further north, towards the Ghôr, I collected *Eremobium lineare, Del.; Monsonia nivea, Dcne.; Anastatica hierochuntina, Linn.* (' Rose of Jericho'); *Astragalus Forskahlii, Boiss.; A. acinaciferus, Boiss.; Rhamnus, sp.* (?) ; *Carthamus glaucus, M.B.; Androcymbium palæstinum, Baker.; Allium Sinaiticum, Boiss.; Aristida ciliata, Desf.; A. Plumosa, Linn.; Panicum turgidum, Forsk.*, with the usual desert species.

The most noticeable feature in the animal life in the northern half of the 'Arabah has been already mentioned. I allude to the extraordinary abundance of small holes and burrows in stone and gravelly sand. The riddled surface reminded me forcibly of the lemming haunts of Discovery Bay, in lat. 81° 45' north, where, however, all were due to one species, with the exception of those of a larger rodent, the stoat, who preyed upon the lemmings. One would expect to find a carnivorous rodent subsisting on the abundant supplies here also, but none such has been as yet discovered. The holes in Wâdy 'Arabah vary from small ant-holes and lizard caches to those of rabbit-holes, and one or two fox-holes (?) were also observed. Tracks of various sizes also abound. Jerboas, porcupine mice, gerbilles, and sand-rats (Psammomys) are the groups represented, of which it is very difficult to secure specimens during a hurried march like ours. Canon Tristram, however, enumerates a considerable variety. One which I trapped here, *Gerbillus erythrurus, Gr*, was sand-coloured, and the size of a large rat, and is now in the British Museum. It does not appear in Canon

GERBILLUS ERYTHRURUS, GR
PP. 44, 238.

A. ZAMENIS CLIFFORDII, SCHLEG. PP. 41, 210.
B. ZAMENIS ELEGANTISSIMUS, GUNTH. PP. 28, 209.
SCALE, ½ NAT. SIZE.

Tristram's work. This gerbille is a wide-spread desert form, from Candahar to Algiers. The holes of this species, and some others, are surrounded outside, besides being well supplied inside, with little heaps of chopped fragments of plants, leaves, seeds, and other remnants of vegetation. Ant-roads are also conspicuous, about an inch wide, and firmly and smoothly pressed down.

Porcupine quills and decomposed remains of hedgehogs were several times picked up in the north end of the 'Arabah.

At 'Ayûn Buweirideh sub-fossil shells were obtained in marl deposits at about 1,400 feet above the level of the Dead Sea, or about 100 feet above sea-level. Two of these, *Melania tuberculata, Mull. ; Melanopsis Saulcyi, Bourg.*, have been figured by Professor Hull at page 100 in his work already referred to. I gathered besides these *Melanopsis buccinoidea, Oliv.*, and *M. eremita, Trist.* These are fluviatile or lacustrine species, and are all found still living round the Dead Sea in various streams and springs. The last-mentioned species is very rare, and I did not find it alive, but Canon Tristram discovered it at the south-western Ghôr. These marls, in the opinion of geologists, are remaining deposits of an ancient lake or inland sea, of which the Dead Sea is all that now exists. From where we now stood to near the source of the Jordan, about 225 miles northwards, must have been a continuous sheet of water in (geologically speaking) tolerably recent times.

Lower marls are very characteristic at an average level of 600 feet above the present level of the Dead Sea. I searched these marls for similar remains in many places, but always found them absolutely barren in records of the past, and very rarely inhabited by any existing life, vegetable or animal. Trunks of palms, floated to and then embedded in these marls at the base of Jebel Usdum, form no exception ; since these may have been drifted thither in times which are as yesterday compared with the 'middle marls.' The upper marls are fairly vegetated with the existing flora. The natural conclusion would be that the ancient sea, at first harbouring fresh-water inhabitants, became reduced by a long process of evaporation, or some other cause, to about a mean height between its present and its earliest level, and that it was already so salt that it was almost if not quite uninhabitable.

At this height, judging from the extent of the middle marls, the waters must have remained stationary for a very considerable period, while most of the upper marls became converted into the lower formation by a long process of denudation. From the latter elevation to the present the subsidence has no doubt been very recent, and is still continuing. The most recent deposits of the Dead Sea are of course perfectly barren, except of mixed drift, or where these have been converted into marshes or fertilized by the few small fresh-water streams.

But I anticipate, in my anxiety to get down to the fertile Ghôr es Safieh.

At 'Ayûn Buweirideh a small flock of pintail grouse circled round the wells, but I failed to obtain a specimen. Subsequently I recognised the note, and obtained the bird, *Pterocles senegalensis, Linn.*, at Bîr es Sebâ. Its call is very peculiar, recalling the strange utterance of the Manx shearwater.

On the night of the 14th we were visited with a thunderstorm and a tremendous downpour of rain. Rain had also fallen on December 3, the day we left 'Akabah; this was our total from Cairo to the Dead Sea. The thunder on the 14th was grand, and continuous for about three-quarters of an hour. Lightning flashed at about every five seconds.

CHAPTER IX.

SOUTH END OF THE DEAD SEA.

On December 16 we obtained our first view of the Dead Sea, and descended to the plain at its southern extremity. The whole depression in which the Dead Sea lies, 1,300 feet below sea-level at its surface, is called the 'Ghôr,' or 'Hollow.' On the first night we camped in the Ghôr el Feifeh, and from the 17th to the 26th inclusive we were detained at the Ghôr es Safieh while waiting for means of transport from Jerusalem.

This enforced delay in so unique a locality was to me a most fortunate circumstance. Previous visitors do not appear to have obtained more than a hurried peep at the Ghôr es Safieh. The difficulties arise from the hostile character of the adjoining tribes of Arabs, who are constantly engaged in predatory warfare, the Ghôr es Safieh being very frequently the scene of their conflicts. Our imaginations were kept excited by continual reports and warnings of those terrible Kerak Sheikhs, Huwaytats, and others who were about to demolish us. I had also read and heard much of the impossibility of doing any good exploring work where an escort is always necessary, and where the Bedawîn were bent on plundering unwary strangers. However, day after day I followed the bent of my inclinations, frequently alone, climbing the eastern hills, searching the jungles and marshes, and collecting birds and plants without ever receiving the smallest annoyance.

The Ghôr es Safieh, where we spent ten days, lies at the south-eastern end of the Dead Sea, about 1,250 feet below the level of the Mediterranean. It is watered by the Garahy River as the Feifeh is by the Tufileh, both descending from the eastern highlands. Between these two oases there is a strip of desert. Both these streams were well supplied with water during our visit, and I understood from the Arabs that the Garahy at least was unfailing. The latter is called also El Ahsi, Hessi, and Safi, and the Nahr el Hussein. Smith's 'Ancient Atlas' calls it the Brook Zered. It is distributed into numerous smaller watercourses for purposes of irrigation by the cultivating Ghawarniheh Arabs, by whose tented village we were encamped. There is another smaller village, called, I believe, El Feifeh, of which we obtained a passing view.

The whole distance from the base of the sudden descent from the barren white marls into the plain is about ten miles to the Dead Sea. The Ghôr es Safieh is about three to four miles wide. The upper Ghôr of El Feifeh is, as I have said, cut off from the lower by a strip of desert, an unwatered patch of sand-dunes and Salsolaceæ. On the east the Ghôr is bounded by the highlands of Moab, and on the west by the briny, muddy, barren bed of the Tufileh. Steep marl banks, a couple of hundred feet high, enclose it on the south, while northwards it gradually becomes salter and swampier, with a diminishing vegetation to the lifeless margin of the Dead Sea.

On the Moab cliffs, as also on the Judæan to the west, the lower declivities are flanked in many places with saline white marls to an upper limit of 650 feet. These marls are absolutely barren *in situ*, but they are fast being washed down by aqueous denudation, and thus purified they are scattered by irrigation over the Ghôr. A minute beetle, of the genus *Galbella*, was a slight exception to this barrenness, which is of course interrupted in the beds and by the margins of the occasional watercourses. This new species, whose description will subsequently be given, is most nearly allied to *G. beccari*, *Gest.*, of Abyssinia.

The upper Ghôr is by no means so fertile as that watered by the larger and more northern stream. The latter issues, with a south-westerly direction, from a narrow cleft, or 'sîk,' in the red sandstone, by which I penetrated for a few miles into that desolate country. The river is here confined to the base of the sharply cut cleft, and confers no fertility on the unaltered marls above. This cleft is 50 to 150 feet in depth, or more, and the period required for its formation must place the marls above at a high antiquity. It should be borne in mind, however, that the water-supply is probably now at its minimum, and the means of erosion were formerly much greater. The bed of this stream was in places absolutely dangerous, from a curious cause. The sides being vertical, there was no upward escape, and the bed of the stream was so deeply clogged with the soft moving mass of silted fine mud that, although there was not more than 18 inches of water, I was compelled, and with difficulty, to retrace my course. As usual when anything risky was attempted, my native deserted me. At its embouchure from the cleft this remarkable stream passes through the lower gravel and shingle deposits, which form the basement of the marls.

On this occasion, when crossing the marls above, I came suddenly upon three ibexes. They whistled or snorted like Highland sheep. I let fly ball-cartridge from my fowling-piece, but missed them. My shots attracted some wild and villainous-looking mountaineers, who followed me to camp that night, where I first became aware of their existence. They could not make themselves understood, but I fancy wished to know if they should hunt the 'beden.' Almost immediately after I lost sight of the ibexes I came across some very interesting and rather extensive ruins of apparently great antiquity. I brought the whole of our party to the spot the following day. The ruins will be found planned and described in Professor Hull's work, at page 121, and again in Major Kitchener's Appendix to the same, at page 216. I leave it to future explorers to identify this site with the ancient Gomorrah.

The following observations were obtained from Sheikh Seyd, of the Ghawarniheh, with regard to the Ghôr:

'Rain generally falls on about ten or twelve days of the year, usually during December and January. Some years there is none. Much more is seen on the highlands on either side, which does not reach the Ghôr.

'They grow wheat, barley, oats, dhourra (sorghum), indigo (one sort), tobacco, and Indian corn.

'Wheat, barley, and dhourra are sown in January; Indian corn in March. Tobacco is sown in January. Indigo is sown in March. They grow some white grapes on trellises. They do not know henna (Lawsonia). Zukkum (Balanites) is common, but made no use of. Mallow is boiled and eaten. Osher (Calotropis) is given to women when barren, or to procure milk, the milk of the bush being taken. Water-melons and cucumbers are cultivated. Of the fruit of the Salvadora (arak) they make a sort of treacle or sweet mixture. Never heard it called "Khardal"; Khardal is mustard, but they have none.

'They (the Ghawarniheh) mostly leave the Ghôr and go up to the hill country in the hottest weather. Snakes and insects are very bad and very numerous in the Ghôr at that season.'

My inquiries about Salvadora were made relative to its claims to being the tree of the mustard-seed parable. I could get no corroboration from these Bedawîn of this view, first put forward by Irby and Mangles, who are not, however, responsible for the statement that it is called 'Khardal' (mustard), nor do they say, as has been misquoted, that they found the 'Ghorneys' using it as mustard. The theory has not, in fact, 'a leg to stand on.'

Mr. Merrill, U.S. Consul at Jerusalem, has kindly made inquiries for me as to the origin of the seed sown by the Arabs. He informs me they save it from year to year, but if they should run short they obtain supplies from Jerusalem. It is to the Mediterranean sea-board westwards, therefore, we must look for the home of any suspicious weeds of cultivation in the Ghôr; and those which are not natives of this region may perhaps be held less open to the question as to their being indigenous in the Ghôr.

No sooner has the river Hessi issued from its unfruitful ravine than the scene changes as if by magic. As it moistens the plain, an extensive growth of bushy, low-sized trees almost covers the district.

In the upper Ghôr these are densely tangled and matted, almost to the exclusion of other growth, and afford shelter for multitudes of birds. In the lower Ghôr the trees are more scattered, often, no doubt, in the more peopled district, from being consumed for firing, and thinned to admit of pasturage and cultivation. These trees are chiefly Acacias (three sorts), Salvadora, Zizyphus, and Balanites. There is also a Rhamnus not unfrequent, and Mr. Lowne mentions Moringa aptera. This latter writer misquotes the authors (Irby and Mangles), whom he criticizes, when he ascribes to them the remark that the oasis contained 'an almost infinite variety of shrubs and bushes.' Their words are: 'The variety of bushes and wild plants became very great,' a phrase which is well within the bounds of the reality.*

Of those trees the Salvadora is the most abundant, and usually occupies a slightly lower region than the Acacias. It grows in clumps, several stems arising together, branching at once, and all combining to form a single tree. It is very leafy above, with small entire leathery leaves; below it displays a labyrinth of grayish branches. The flowers and fruit are small and numerous. It attains a height of about 20 feet, a stray branch reaching to 25 or 30 feet. The Balanites (Zukkum) is usually a smaller tree, and was in full fruit. This is green and wrinkled, somewhat like that of a walnut. The leaves are few and small. The Zizyphus is the well-known sidr or thorn of the Arabs, the 'dôm' when reaching a large size. Its branches, strewed in lines along the ground, form the fences to protect the grain from cattle.

As the plain slowly lowers to the Dead Sea, becoming at the same time gradually moister, the vegetation changes. The above species decrease in the number of individuals. Tamarisks, Osher, Salsolas, Prosopis, and Atriplices take their place in abundance. Of these, the Osher (*Calotropis procera*) is the most remarkable. It is somewhat like a gigantic small-leaved cabbage bush, cactus-like, and with the bark of a cork-tree—utterly strange-looking to European eyes. Its fruit, the size of a large apple, is full of silk and air, and is probably

* I quote from Murray's edition in the Colonial and Home Library, vol. iv., p. 108, ed. 1884.

to be identified with the 'apples of the Dead Sea.' The drawing of these 'trees that beren fulle faire apples, and faire of colour to beholde,' by Sir John Maundeville, is by no means unlike the Osher. If the early traveller's figure stands for any real thing, it is probably for this bush, which here attains a remarkable size. Of it the writers already quoted say: 'We were here (Ghôr es Safieh) surprised to see for the first time the Osher plant, grown to the stature of a tree, its trunk measuring in many instances 2 feet or more in circumference, and the boughs at least 15 feet in length, a size which far exceeded any we saw in Nubia; the fruit also was larger and in greater quantity.' This remark is interesting in connection with Captain Burton's, that the Osher in South Midian is 'a tree, not a shrub' ('Land of Midian,' ii. 206), as though the plant was more at home in the Eastern continent. The *Castor-oil* (*Ricinus communis*) is also very conspicuous and large (20 to 25 feet), chiefly in the same localities as the Osher. Other bushes are the leafless Leptadenia pyrotechnica, and the poplar, *Populus euphratica*. All these were seen in the Ghôr el Feifeh also. A tree of the latter, about 50 feet high, near the Dead Sea, is, I think, the largest tree in the whole Ghôr. Oleanders and Osiers are confined to the embouchures of the stream from the mountains or farther up.

As we approach the Dead Sea, occasional swamps produced jungles of various late grasses, chiefly *Arundo Phragmites* (*P. gigantea, J. Gay.*); *Erianthus Ravennæ, P. de B.*, and *Imperata cylindrica, P. de B.*, mixed with several Cyperaceæ, of which the most interesting were *C. eleusinoides, Kunth.*, and sparingly, I believe, *C. papyrus, Linn.* Salter patches are given up to *Juncus maritimus* and *Eragrostis cynosuroides, Retz.* The former (var. arabica) was from 4 to 7 feet high. Tamarisks, Suædas, Salsolas, Salicornia, and Atriplices are the last to fail. Tamarisk, Salicornia herbacea, and Ruppia not in flower, probably *R. spiralis, L'Her.*, were the very last; the former all along the inner margin, the latter two where the mud of the sea is in union with that of the Tufileh estuary. The latter two encroach downwards upon the forbidden area here, from salt swamps to those which are too salt, as they do upwards in our own country, from salt swamps up fresher estuaries until they meet those which are too fresh.

A brief space, 50 yards or more, varying with the slope and the fulness of the basin, is barren saline mud or sand. This foreshore is at other seasons under water, and all which is liable to be submerged is barren, except in the two instances above mentioned on the Tufileh mud.

An interesting assemblage of sea plants is congregated around the Dead Sea. These are *Sonchus maritimus, Linn.; Inula crithmoides, Linn.; Lotus tenuifolius, Rchb. (Lythrum hyssopifolium, Linn.); Salicornia herbacea, Linn.; Salsola; Suæda; Atriplices; Scirpus maritimus, Linn.; Fimbristylis dichotoma, Rottb.; Juncus maritimus, Linn.*, and *Ruppia, sp.* (?) (*R. spiralis, L'Her.* ?). Some of these at first sight will hardly fail to impress the observer with the idea that the vegetation must recently have undergone distinct maritime conditions; but a little reflection will show that the visits of aquatic birds, and the present suitability of the circumstances, suffice to explain their presence. Moreover, the most conspicuous are of the easily-diffused pappus-bearing compositæ.

Several of the most interesting species were obtained by penetrating into the jungles in all directions. In the very heart of these, Cynanchum acutum was abundant, trailing convolvulus-like about the reeds. These jungles, and along the banks of the stream, were my best hunting grounds.

The luxuriance of some familiar British aquatic plants may be alluded to. The sea rush, as already mentioned, reaches 7 feet in height, Inula crithmoides 4 to 7 feet, and Lycopus europæus 5 to 6 feet in height, while gigantic plants of Lythrum salicaria had reached a height of 14 feet.

One of my most interesting 'finds' was that of a handsome acacia, *A. læta, Br.*, in the Ghôr. This species has not been recorded north of Syene (Assouan) in Upper Egypt, seven degrees farther south. There were several trees of this very distinct species, which is much larger and better furnished than the other acacias met with. An Arab to whom I silently pointed out one of this species at once exclaimed 'Sont,' and proceeded to show me the difference in its leaves and fruit from that of a Seyal, its neighbour. At 'Akabah an Arab called a large A. tortilis 'Sunt.' It is an Egyptian name, but never applied to the Seyal.

A few other remarkable species not noticed by previous botanists in Palestine may be mentioned: *Cocculus Læba, D.C.; Sclerocephalus arabicus, Boiss.; Zygophyllum simplex, Linn.; Indigofera paucifolia, Del.; Rhynchosia minima*, D.C.; *Trianthema pentandra*, Linn.; *Eclipta alba*, Linn.; *Pentatropis spiralis*, R. Br.; *Salsolaceæ* (several); *Digera arvensis*, Forsk.; *Boerhavia verticillata*, Poir.; *B. repens*, Linn.; *Euphorbia ægyptiaca*, Boiss.; *Cyperus eleusinoides*, Kunth., and some others. Several of these are distinctly tropical, and add to that most interesting group of those plants already known to inhabit the 'sultry Ghôr.'

I gathered altogether at the southern end of the Dead Sea about 225 identifiable species of flowering plants. The total there may reach 300. Many annuals and Mediterranean spring plants, especially of the Leguminous and Cruciferous orders, were still in a young condition.

I defer a fuller analysis for the present, merely remarking that the flora of the Ghôr, a unique locality, is even more interesting, and that in no mean degree, than it has hitherto been shown to be.

The Ghôr has been visited by two competent botanists, Messrs. B. T. Lowne in 1864, and W. Amherst Hayne in 1872, both in Canon Tristram's company. These gentlemen have, however, hardly dealt with the oasis of Es Safieh. Mr. Hayne's essay, appended to Canon Tristram's 'Land of Moab,' is only enough to make a botanist wish for more of it, while Mr. Lowne's valuable paper, published by the Linnean Society, deals with the south-western extremity of the Ghôr, two dry desert wâdies whose flora is the northern wave from Sinai and the 'Arabah.

Although devoid of life, the sandy beach of the Dead Sea mentioned above was full of interest. On it were strewed salted remnants of a variety of insects, beetles, spiders, locusts, and seeds which had been floated from the Ghôr by the rivers and promptly killed and cast ashore. Several of these were identifiable, although of no value as specimens. A better collection in the same place was that of shells. In some places these were thickly strewn, and I went through these natural museums with the greatest care, obtaining thus several varieties not previously found in Palestine. Amongst these are *Planorbis albus*, Mull.; *Limnæa peregra*, Desf.; *Physa contorta*, Mich.; *Achatina (Cionella) brondeli*, Bourg.; *Ferrusacia thamnophila*, Bourg., and a new species of Bulimus.

The tamarisks near this were inhabited by a species of ant. These make their home, in parties of 20 or 30, in a sort of purse of vegetable matter, made out of scraps triturated together and worked into a smooth papery lining. The species is *Polyrhachis seminiger*, Mayr., belonging to a tropical, chiefly Indian, genus. Multitudes of little fishes, *Cyprinodon dispar, Rupp.*, as mentioned by Tristram, were seen in the salt pools close by.

Although my visit was too early for many species of plants, yet on my first day in the Feifeh I found at once numerous kinds not seen in Sinai, of which a good many were both in flower and fruit. These must flower continuously, or with a very brief respite; others, chiefly European and Mediterranean species, were rapidly advancing to the flowering stage during our sojourn in the Ghôr.

A good number of Sinai species occur in the Ghôr. An effect of the moister climate on some of the woolly desert plants was noticeable. These became very perceptibly less so in the Ghôr. Pulicaria undulata, P. arabica, Tribulus terrestris, Verbascum sinuatum, may be instanced. Possibly the salinity of the atmosphere assists in this; the tendency of plants to become glabrous by the seaside is familiar. On the other hand, excessive dryness appears to provoke pubescence in plants, as well as other striking qualities of pungent odours, gummy exudations, and conversion of leaves to spines, all of which we may expect to find diminished if the species can accommodate itself to moister conditions.

I have hitherto spoken almost entirely of the plants. The district is of as great interest in other branches of natural history. Canon Tristram's various works have made this fact familiar. My prolonged stay at an unusual season must indeed be my excuse for trespassing on a subject he has made so peculiarly his own.

The Ghôr swarmed with birds. About forty species were observed, of which, with two or three exceptions, specimens were obtained. Some, especially doves of two species, and bulbuls of the sort already met, were extraordinarily abundant. The doves were the Indian collared turtle, *Turtur risorius, Linn.*, and a smaller beautifully bronzed species, *T. senegalensis, Linn.*

On the Dead Sea mud, redshanks, lapwings, and sandpipers flitted and fed, but they were confined to those parts of the margin which were tempered by fresh water. Snipe, water-rails, and ducks of British sorts were frequently met with. Marsh sparrows in great flocks also kept near the shore. Buntings and larks of three sorts were in vast numbers throughout the stubbles of maize. The two desert partridges occurred on the margins of the Ghôr, where also the thick-knee was shot. Shrikes, 'boomey' owls, marsh harriers, buzzards, sparrowhawks, and kestrels were all noted. The mellow, loud whistle of Tristram's grakle frequently caught the ear, as did also the excessively discordant craking note of the Smyrna kingfisher. The beautiful little sunbird and the gaudy blue-throated robin were about equally common, the former usually frequenting those acacias which gave support to the handsome Loranthus. Several other warblers were observed, but for most of these, as well as the swifts and others, the season was too early. On the upper ground at the edge of the Ghôr several pairs of desert chats of two or three kinds might be always studied, and the impression the Ghôr gave me was that many migratory species of Palestine who ought to travel south from the Jerusalem plateau in winter found here a conveniently close and sufficiently warm retreat, which they utilize in vast numbers.

Burrowing animals still gave evidence of their abundance. Traps set for these were, I believe, appropriated by Bedawîn lads, for I could never rediscover them. The traps were strong, and I trust they snapped on their meddlesome fingers. Jackals kept up their high-pitched scream throughout the night. Bedawîn, bantams, jackals, and jackasses have all peculiarly high notes in the Ghôr. They howl together in a shrill minor key, chiefly when they ought to be asleep.

Fresh boar tracks were always visible; on one or two occasions I heard the animals crashing in the jungle close ahead of me. Ibexes were seen in the ravines close by.

There are many cattle scattered through the Ghôr. These are chiefly small pretty black animals with white faces, somewhat like the Highland breed, while goat-like sheep and sheep-like goats, with ears hanging 6 inches below their snouts, are herded evening and morning. Donkeys are more numerous than ponies; there are very few of the latter in the possession of the much-molested and peaceful Ghawarniheh.

The Bedawîn supplied us with poor milk and very small eggs.

Insect life had as yet hardly awakened. About half a dozen species of butterflies were observed, of which some were Ethiopian forms. Scorpions were still torpid. Molluscs, except fluviatile, were scarce, while Batrachians and Reptilia might have been almost non-existent, with the exception of the Lacertidæ.

A very nimble fresh-water or rather marsh crab was very abundant. To this animal was due the multitude of burrows amongst the tufts of *Juncus maritimus* near the Dead Sea. Twice I saw them disappear with incredible swiftness into these holes, which were of various sizes, and of so great a depth or length I could not usually dig them out. Several that I did dig out were blind or empty, and at first these holes puzzled me beyond measure. The total absence of tracks or pads leading to them arrested my attention, while their widely different sizes, both in length and diameter, suggested something altogether new. Those crabs I obtained were by means of the Bedawîn lads. The carapace of the biggest was about 5 inches by 3. They are gray in the young state, but attain a reddish tint when full grown. The species is *Telphusa (Potamophilon) fluviatilis, Savign.* One was killed in our camp, showing that they ramble at night away from water or marshy places. This crab extends through Egypt to Algiers, and occurs also, I believe, farther east than Palestine.

At the time of our visit the mean diurnal temperature was about 50° Fahr. There is no universal check to vegetation in the Ghôr. Acacias, Osher, Castor-oil, Loranthus, Salvadora, species of Abutilon, Zizyphus, and Balanites were bearing fruit and flower now in the coldest season in true tropical fashion.

Before we left, the sun was just beginning to 'braird the lea,' and there was a delicate hue of green perceptible across the ill-tilled soil.

The river, Seil Garahy, alias Hessi, was well filled with water, and on several occasions we enjoyed a swim down the swift deep rushes at the inner edge of the plain. Irby and Mangles, I think, found this river dry on their return journey from Petra.

Before bidding farewell to the Ghôr I should mention one striking peculiarity in its flora. I allude to the great number of species compared with the number of individuals. If those few gregarious kinds (chiefly trees, grasses, and shrubs) already mentioned be eliminated, the remaining sorts would very often depend on a few plants for their claim to a place in the list. Hence a brief visit may give rise to many omissions.

CHAPTER X.

GHÔR ES SAFIEH TO GAZA.

On December 27 we finally struck tents and left our camp in the Ghôr es Safieh. As we passed westward near the south end of the Dead Sea some interesting features were observed. The waters vary in their surface level about 3 feet between the brief wet period and the minimum level. During our visit they stood at a low level, and the drift of timber and terrestrial shells showed an upper margin at a uniform height in several places. Where the shore slopes very gradually, as in most places round the southern end, this variation in depth is sufficient to leave a wide space of foreshore uncovered. This was very noticeable during our journey along the base of Jebel Usdum, at the south-west corner of the Dead Sea. The water was there about 600 yards from the line of drift. Inside this was the usually traversed track along the base of Jebel Usdum, and above, about 7 vertical feet higher than the present high-water drift, was an older well-marked margin, looking very recent and pointing to a still continuing evaporation of its waters in excess of the supply.

Logs of palm-trees frequently marked these margins, and these were seen embedded in a drifted position in the marls of Jebel Usdum as much as 27 feet above the highest level now attained by the waters of the sea. Palm-tree trunks were also seen along the river Tufileh in the Ghôr el Feifeh and lower about its estuary. These were probably, from their appearance, torn out of its bed during a flood in a semi-fossilized condition. Thus the subsidence of this sea has continued and is continuing, and earlier deposits are being continually carried down to form more recent ones and to fill up the cavity. Most parts of the Dead Sea south of the Lisân are very shallow. In two places, when looking for a swim, abreast of Jebel Usdum and north from the Ghôr es Safieh, I waded out several hundred yards without getting water above my knees, and the water, like that at the mouth of the Jordan at the other end, is usually turbid. The work of reclamation steadily proceeds, and as the sea is known to be of very considerable depth (200 fathoms) in other places there is abundant room for the inflowing sediment.

Of Jebel Usdum I have given a description to Professor Hull, which has appeared in his account of our expedition. It proved, as it looked, to be of little botanical interest, and I should not have climbed it had I not seen it stated in several places that it was inaccessible. The plants found on its upper portion, 650 feet above the Dead Sea, were very few, the whole being a bare flat with a slight central ridge of barren marl—the cap of the central core of rock-salt. A couple of solitary tamarisks occurred and several Salsolaceæ. The latter were *Noæa spinosissima, Moq.; Atriplex alexandrina, Boiss.; Salsola rigida, Pall.*, var. *tenuifolia.; S. tetragona, Del.; S. fœtida, Del.*, and *S. inermis, Forsk.* The 'mountain of salt' is, in fact, well characterized by this order. Several of the above are additions to the flora of Palestine. On the western slope a few desert species of the ordinary and familiar types were collected, and these gradually increased to the base at the Mahauwât Wâdy, whose flora has been already the subject of a special paper by Mr. Lowne. This writer gathered here, and in the neighbouring Wâdy of Zuweirah, eighty-two flowering species, chiefly of the desert sorts. These are all, or almost all, either Sinaitic or occur in the Wâdy 'Arabah.

Leptadenia pyrotechnica, Forsk., and *Ochradenus baccatus, Del.*, grow to a large size here. The latter was about 15 feet high, close to the Dead Sea, at the confluence of these two wâdies. *Zilla myagroides, Forsk.*, was here in flower, bearing a pretty little blossom like our Cakile maritima.

During the ascent of Wâdy Zuweirah to the plain of South Judæa the following fresh species were collected: *Notoceras canariense, R. Br.; Enarthrocarpus lyratus, D.C.; Zollikoferia, sp.* (?) (*Z. stenocephala, Boiss.*?); *Lithospermum tenuiflorum, Linn.; Heliotropium rotundifolium, Sieb.; Ballota undulata, Fres.; Arnebia linearifolia, D.C.*, and *Plantago Loefflingii, Linn.* A large bulb, *Urginea Scilla, Stein.* (?), now only in leaf, marks well the transition stage from the Ghôr flora to that of the Judæan wilderness. Desert species, as Fagonia, Zygophylla, Retama, Acaciæ, Resedaceæ, Cucumis, Microrhynchus, Dæmia, Ærua, Forskahlea, and others were here for the most part taken leave of. These ascended perhaps a third part of the climb, several ceasing at about the old Saracenic Fort. Upwards, and on the Judæan plain, a great change takes place. We found ourselves ere long on rich land arousing itself to a spring growth, although the most inclement season was not yet reached.

FAUNA AND FLORA OF SINAI, PETRA, AND WADY 'ARABAH. 59

The need of water is of course everywhere apparent. Withered remains are scarcer than in the desert, and the ground is often bare for considerable spaces, or with a few early patches of species to be presently mentioned. It becomes difficult to recall the existence of the contiguous Ghôr flora, with its perennial luxuriance. Hardly a bush and no trees are observed to break the monotony. Travelling still westwards, evidences of cultivation, that is to say of the soil being ' scratched ' and sown, appear. Soon after Bîr es Seb'a, two days from the Ghôr, we find ourselves amongst softly-swelling downs covered with sowers and ploughers, but otherwise monotonous in aspect, as the cretaceous limestone formation usually is.

The species first observed at the head of Wâdy Zuweirah and upwards to Bîr es Sebâ were numerous, many of them spring Mediterranean species just opening their flowers. The following were conspicuous: *Carrichtera Vellæ*, D.C.; *Biscutella Columnæ*, *Ten.*; *Enarthrocarpus lyratus*, *Del.* (?); *Silene dichotoma*, *Ehr.*; *S. Hussoni*, *Boiss.*; *Helianthemum Kahiricum*, *Del.*; *Astragalus sanctus*, *Boiss.*; *A. alexandrinus*, *Boiss.*; *Erodium cicutarium*, *Linn.*; *Senecio coronopifolius*, *Del.*; *Scorzonera lanata*, *M.B.*; *Calendula arvensis*, *Linn.*; *Achillea santolina*, *Linn.*; *Anchusa Milleri*, *Willd.*; *Cyclamen latifolium*, *Sibth.*; *Ajuga Iva*, *Schreb.*; *Satureia cuneifolia*, *Ten.*; *Marrubium alysson*, *Linn.*; *Salvia verbenaca*, *Linn.*; *S. controversa*, *Ten.*; *S. ægyptiaca*, *Linn.*; *Eremostachys laciniata*, *Linn.* (in leaf only); *Paronychia argentea*, *Lam.*, and *Urginea undulata*, *Steinh.* (?). Several of these are pretty little bright-flowered yellow and blue annuals.

We were now travelling on horseback, and I had no longer the same facilities for botanizing. The pace was usually too fast. My method was to keep well ahead till I reached some inviting point, and then dismount and botanize, usually holding a rein across my arm. The result was that I was usually left far behind, or in hot pursuit of the party. Sometimes I lost my way altogether. It would have needed a botanical circus rider to get on and off his horse with comfort as fast as new flowers occurred.

Several mosses and lichens were gathered on this march. The mosses were *Tortula muralis*, *Linn.*; *Bryum atropurpureum*, *W. et M.*, and a Hepatica, *Riccia lamellosa*, *Raddi*. The mosses are both British species.

8—2

In animal life, gazelles, mole-rats, *Spalax typhlus, Pall.*, and sand-rats, *Psammomys obesus, Rupp.*, appeared to be the most abundant. I captured examples of the latter two, which are now in the British Museum.

The mole-rat, the Asiatic representative of the English mole, though of a very different family, is a strangely ugly little animal with long protuberant teeth. Mr. Armstrong showed me a ready way of obtaining specimens, which at first sight appeared to be hopeless. His plan was to watch the freshly up-lifted heaps of soil which are raised in line at short intervals, and notice the direction the animal is burrowing in by the relative freshness of the heaps. Soon a slight movement will be observed in the freshest heap or beyond it, and on firing a charge into the ground at once, the gun about a foot from a point a few inches ahead of the moving place, the animal will be stunned, and may be at once dug out, probably alive. I tried this plan twice successfully.

A buff-coloured snake, about 3 feet long, *Zamenis atrovirens, Gray.*, was killed in the neighbourhood of Tell Abu Hareireh. Geckos and toads were also captured. A brown and gray fox (Vulpes nilotica, ?) was seen near Bîr es Sebâ. Laurence shot a fine wild cat (*Felis maniculata, Rupp.*) in a gulley near Tell Abu Hareireh. It measured 2 feet 8 inches from the tip of the nose to the tip of the tail, the tail itself being 1 foot. It was of a grayish-brown colour, brindled with sandy brown across the back and down the sides. The tip of the tail was ringed with black. This is supposed to be the cat found embalmed in Egyptian monuments. It is found along the Nile, and as far south as Abyssinia.

I spent as much time as I could in digging up bulbs. Of these there were several identifiable species, as *Xiphion palæstinum, Baker.*, a dwarf sweet iris, with large flowers in tints of buff and French gray. *Colchicum montanum, Linn.*, occurred in the greatest abundance, white or pale mauve, and was very beautiful. *Urginea Scilla, Sternih.*, and *Asphodelus ramosus, Linn.*, were most abundant, increasing westwards to Gaza. *Bellevallia flexuosa, Boiss.*, and *Ornithogalum umbellatum, Linn.*, also frequently appeared.

About Bîr es Sebâ the birds observed were cranes, black and white storks, buzzards and kites, trumpeter bullfinches, pintail grouse, Greek partridge, black-headed gulls and lapwings, as well as several desert larks and chats. The technical names of these species will subsequently be enumerated. The trumpeting of the crane was heard frequently, usually at night.

At Tell el Milh, in a swamp, a flock of teal was flushed, and a number of the black or Sardinian starlings came to roost in the rushes. Their note is different from that of our species. A snipe handsomely marked with white, as in flight, with a rich brown back, and showing vivid green tints also on the upper surface, was unfortunately missed. It uttered a peculiar quacking cry, and I had several good views of it. There were three or four birds in the marsh, and I have no doubt it was the painted snipe, *Rhynchæa capensis, Linn.*, which has not previously been known to inhabit Palestine. It is a widely-spread species in Africa.

The Cyclamen and the Colchicum are constantly exciting our admiration. In the marsh just mentioned *Spergularia marginata, Koch.; Cyperus longus, Linn.*, and *C. lævigatus, Linn.*, var. *junciformis*, were collected.

A feature noticed by all travellers is the abundance of snails on the small shrubs, chiefly on Anabasis articulata, Boiss. The commonest of these was perhaps *Helix Seetzeni, Koch.*, but I also gathered *H. joppensis, Rottb.; H. syriaca, Ehr.; H. protea, Zugl.; H. vestalis, Pass.; H. tuberculosa, Conrad.; H. candidissima, Drap.; H. Boissieri, Charp.*, and *H. cavata, Mouss.* H. cavata and H. Boissieri are the finest of these species in size, the latter being a heavy solid-shelled sort. H. tuberculosa is trochiform, or top-shaped. This species and his flattened brother, *H. ledereri, Pfr.*, gathered between Gaza and Jaffa, are both scarce. They are the prettiest, being delicately mitred and foliated at the whorls.

The black-headed gulls, and no doubt others of the birds, subsist on these molluscs.

Continual evidence of wild boars occurred, and some of our party had the good luck to obtain a sight of a 'sounder,' or family party. They seem to feed chiefly on the bulbs, of which some large kinds are marvellously plentiful. An Urginea (probably *U. undulata*) was sought after especially, so that it was with difficulty roots which they had not mashed were obtained to bring home. It has since flowered, and in the absence of leaves is doubtfully referred to this species by Mr. Baker.

Urginea Scilla covers the ground for miles, and grows sometimes to the exclusion of everything else. It appears to be a scourge to the fellahîn. Great heaps of its bulbs, the size of a melon, are often met with, and lines of its growth are commonly left to mark off each cultivator's allotted space. *Asphodelus ramosus, Linn.,* is nearly as common. The brilliant anemone (*A. coronaria, Linn.*), the 'lily of the field,' was picked in flower on the last day of the year. The curious stringy *Thymelæa hirsuta*, whose acquaintance I first made on the shores of Brindisi on the outward journey, is profusely common. Between Bîr es Sebâ and Gaza the species now in growth are almost altogether of the Mediterranean type. A few desert species occur, but chiefly of a Syrian or Mesopotamian character, as *Caylusea canescens, Deverra tortuosa, Alhagi maurorum, Peganum harmala, Citrullus colocynthis, Artemisia herba-alba,* and *Anabasis articulata*.

The universal 'rimth' (Anabasis or Salsola) of the Sinai Bedawîn is called by the Doheriyeh Arabs 'Shegar.' It may be that the Arabs put off inquiries from one whom they perceive to be unlearned in their language with trivial and unmeaning terms; but the results of my short experience would tend to show that little importance can be attached to these local names. Different tribes and places yielded different terms, so that on comparing my collection of Arab plant-names with those given by several other writers, hardly two were identical, or even alike. In the Serbâl district of Sinai, Wâdy Rimthi takes its name from the Anabasis.

The soft note of the trumpeter bullfinch, rising and falling as if borne on the wind, while the bird is concealed on the ground somewhere close by, often arrested my attention. It was impossible to tell whether it was ten yards or ten times that distance away.

Travelling west past Tell Abu Hareireh to Gaza, the following plants occurred in addition to those mentioned already about Bîr es Sebâ: *Malcolmia pulchella, Boiss.; Matthiola humilis, D.C.; Alyssum Libyca, Viv.; Erucaria microcarpa, Boiss.; Capsella byrsa-pastoris, Linn.; Polycarpon succulentum, Del.; Dianthus multipunctatus, Ser.; Silene rigidula, Sibth.; Ononis serrata, Forsk.; Hypericum tetrapterum, Fres., forma; Erodium hirtum, F.; Bupleurum linearifolium, D.C. (?); Carthamus glaucus, M.B.; Thrincia tuberosa, D.C.; Tolpis altissima, Pers.; Scorzonera alexandrina*, Boiss.; *Mandragora officinarum, Linn.;* *Withania somnifera, Linn.;* *Echium plantagineum, Linn.;* *Lamium amplexicaule, Linn.;* *Euphorbia exigua, Linn.;* *Paronychia nivea, D.C.;* *Andropogon hirtus, Linn.*, and *Poa annua, Linn.*

CHAPTER XI.

GAZA TO JAFFA.

AT Gaza we were kept a few days in quarantine by the Turkish authorities. This was not because we were deemed infectious (the idea was absurd), but to levy a tax on our purses. By the prompt interference of Lord Dufferin, British Ambassador at Constantinople, to whom we telegraphed, we were released in four days instead of being confined for a fortnight.

This delay was to me most valuable, as it enabled me to sort my rapidly-made collections of the last few days.

On our last day, having liberty to leave quarantine ground, I gathered a good many species south of Gaza which I had not seen before. Many of these belong to well-known Mediterranean types, but there is still an important admixture of desert and Egyptian forms, belonging to a somewhat more southern group.

Gardens of fruit trees, olive groves, and enclosures hedged by the prickly pear (*Opuntra vulgaris, Linn.*) reached our camp from the inland side. On the leeward we were hemmed in by high sandhills, the vanguard of an ever-advancing column, driven westward by the prevailing winds, which is gradually swallowing up Gaza, old and new, as well as a long belt of coast north and south of it.

Some laborious journeys across this belt of sand, often three or four miles broad, impress them vividly on my memory. They yielded exceedingly few species, being as a rule completely barren. I may mention *Silene succulenta, Forsk.;* *Scrophularia xanthoglossa, Boiss.;* *Euphorbia terracina, Linn.*, which grew well out on the dunes.

These sands are effecting a steady and enormous change along the coast. It is difficult to reach what is left of Ascalon, which remains on an insulated patch of rocky ground by the sea, completely cut off inland. Little of it is left unsmothered. Ashdod is undergoing the same fate. Gaza retreats inland in front of the arenaceous sea, and it is only at intervals, or by ascending some eminence which is rarely met with, that one obtains even a view of the Mediterranean. This was to me a keen disappointment, and I sighed for the reality of a cliff-girt coast like that of my home in north-western Donegal.

In and about the Gaza olive groves several birds familiar at home abounded. Others occurred on the plain hard by. It was refreshing to hear their well-known voices in this strange and inhospitable land. There were English sparrows, swallows, buntings, goldfinches, black redstarts, chaffinches, stonechats, willow-wrens and chiffchaffs, blackbirds, and hooded crows. Other birds seen were Egyptian kites, buzzards (common species), 'boomey' or little southern owl, red-breasted Cairo swallows, pelicans, dunlins, calandra and crested larks, bulbuls, pied chats, and Menetries' wheatear.

At an estuary about four miles south of Gaza, and up a flat wâdy leading to it, I obtained several good plants. This would be capital ground to botanize at a later season. The following are the most interesting: *Brassica Tournefortii, Gou.; Cratægus azarolus, Linn.; Neurada procumbens, Linn.; Ceratonia siliqua, Linn.; Astragalus aleppicus, Boiss.; A. macrocarpus, D.C.* (not in fruit); *Medicago laciniata, All.; Ononis natrix, Linn.,* var. *stenophylla; Anagyris fœtida, Linn.; Acacia albida, Del.; Prosopis stephaniana, Willd.; Xanthium strumarium, Linn.; Artemisia monosperma, Del.; Centaurea araneosa, Boiss.; C. pallescens, Del.; Atractylis prolifera, Boiss.; Linaria Hælava, Forsk.; Anchusa ægyptiaca, Lehm.; Prasium majus, Linn.; Andrachne aspera, Linn.; Ficus sycomorus, Linn.; Ricinus communis, Linn.; Boerhavia verticillata, Poir.; Plantago albicans, Linn.; Euphorbia peploides, Gou.; Emex spinosus, Camp.; Salsola inermis, Forsk.; Cyperus schœnoides, Griseb.; C. rotundus, Linn.; Fimbristylis dichotoma, Rott.,* and *Pennisetum cenchroides, Rich.* Some of these, as the castor-oil, the little anomalous desert Neurada, and the tropical Boerhavia, point to the great heat of Gaza.

The trees about Gaza are chiefly date-palms, olives, sycamore fig, caroub (Ceratonia) or locust-tree, and fig; a very handsome tamarisk (*T. articulata, Vahl.*) reached a height of 30 or 40 feet, with bright green foliage, very refreshing and home-like after the dull gray or lifeless green of the desert. The olives are of enormous age. They usually have unbranched trunks, 2 or 3 feet in height, then perhaps divided, and at 7 or 8 feet the leafy canopy, browzed below to a level height by cattle, begins. The average height of the tree is 20 to 25 or 30 feet. Old trees have often mere shells of their trunks remaining. I measured the two largest I saw, a few miles north of Gaza; their girth was 18 and 20 feet respectively at 2 feet from the ground, a size which was maintained, or very nearly so, till the trunk forked.

At Ascalon, which Laurence and I visited at a gallop just before dark, I gathered *Calycotome villosa, Linn*, in the sands, a pretty yellow shrubby pea-flower. Ascalon is a wilderness of shifting sands. The small space of remaining earth is inhabited by a few Arabs, from whom I got my first Jewish coins. Several pillars of marble and black granite lie about the ruins of the crusading fort, but none are in position.

Frequently dogs with unmistakable traces of jackal parentage were seen along here. I was assured it is by no means uncommon for these animals to interbreed along this part of the Mediterranean seaboard.

The chief crop showing was of lentils. I saw bean-stalks a foot and a half high in the first week of January.

A few of the commonest British plants, as Capsella bursa-pastoris, Silene inflata, Convolvulus arvensis, and Rumex obtusifolius, occur along here.

A handsome tree introduced from the East is very common. It is the Melia azederach, or Pride of India. It is deciduous, and was only bearing fruit, as I saw it, along the enclosures or by the villages. *Lycium europæum, Linn.; Rubia Olivieri, A. Rich.; Ephedra alata, Dcne.; Asparagus aphyllus, Linn.,* and *A. acutifolius, Linn.,* are the larger plants, which help to stop up the gaps in the prickly pear fences.

At Yebnah, and thence to Jaffa, *Narcissus Tazettæ, Linn.,* was in flower. Some damp low-lying patches were white with it. Other species were *Ruta graveolens, Linn.,* Erodium sp. (?) (*E. bryoniæfolium,* ?); *Retama retam, Forsk.* (in flower); *Lithospermum callosum, Linn.; Echiochilon* *fruticosum, Desf.; Thymus capitatus, Linn.; Lavandula stœchas, Linn.,* and *Rhamnus punctata, Boiss.* The Retem broom was in flower, very pretty, white variegated with purple. I found it once previously in blow in the desert.* *Lawsonia alba, Linn.* (henna) was seen several times, but usually here (as at 'Akabah) either in or on the verge of enclosures. No doubt it remains from ancient gardens at Engedi, where it was, I believe, abundant. It is native much farther east.

In the gardens next the hotel at Jaffa were some very interesting plants. I did not learn their history, or who made the collection. Some of the Sinaitic and Dead Sea plants were there—the handsome trailing pea, *Dolichos lablab*, which I found in the Ghôr, a widely cultivated plant in hot countries, but perhaps originally introduced from India. The Sinaitic Gomphocarpos, a milky asclepiad with pods full of silk, one of the most remarkable species in the peninsula, was here also; it differed, however, from the Sinaitic plant in being shrubby and about 6 feet high, while the desert plant averaged from a foot to a foot and a half.

Ricinus communis (the castor-oil); Echaverias, Lavandula stœchas (the handsome purple woolly lavender just mentioned), and quite a collection of Acacias and Mimosas, with oranges, bananas, indiarubber trees, fan-palms, Eucalyptus, Mesembryanthemums, and many others made up a tropical garden which will well repay the traveller's visit. I was peculiarly interested to see my Boucerosia from Mount Hor here, a cactus-like plant, which seems to be a new species. Can it be, like the Dolichos, an ancient weed of cultivation? When we let the mind go back to times of ancient civilization, to the traffic and merchandise of pilgrims, monks, and Bedawîn, of Israelites and Phœnicians, Pharaohs and Ptolemys, Greeks and Romans, Turks and Crusaders, caravans and ships laden with food, with gums, spices, fruits, and wares during the whole history of mankind, we must reflect that many plants we now view as inhabitants, especially those of any economic use, may have hailed originally from remote sources. Speculations of this kind, at

---

* This is the Hebrew 'rothem' or 'rotem,' translated juniper in the Old Testament. The same name (Retama) is applied to a species of a closely-allied genus, the *Spartocytisus nubigenus*, of the middle zone of vegetation of the Peak of Teneriffe, as I learn from Mr. Moseley's 'Notes by a Naturalist on the "Challenger,"' p. 5.

once so uncertain and so unpalatable, had better perhaps not be indulged in. They can only lead to doubt and discussion. Granted that the 'osher' is known by the Bedawin 'Doctrine of Signatures' as a plant of domestic value, may we not theorize as to whether wandering tribes have not carried it from Midian or Nubia to Sinai? from Sinai to its far northern home in the Ghôr? and so with many others.*

The gardens at Jaffa were fully supplied with its own brand of most excellent oranges.

CHAPTER XII.

JERUSALEM.

BETWEEN Ramleh (a few miles from Jaffa) and Jerusalem, during an ascent of over 2,000 feet, many fresh species occurred. The chief change in plant life lay in the great increase of low shrubby vegetation on the limestone hills and terraces. I had little time to botanize, but with hard galloping to make up for delays, I secured several sorts in condition to be studied. An oak, *Quercus coccifera, Linn.*, and the handsome large-leafed arbutus in full flower, *Arbutus andrachne, Linn.*, are two conspicuous trees or bushes characteristic of the rocky regions above the plain of Ramleh. A large daisy, *Bellis sylvestris, Cyr.*, similar except in size to our own Bellis perennis, was in flower. The handsome locust-tree, usually here of only the stature of a bush from being cut for firing like the others, is very frequent. Its rich dark green pinnate foliage is well known to travellers in Southern Europe, where its pods are much used to feed cattle. This is supposed to be the 'locust' of St. John. At Kirjath-jearim a solitary date-palm occurs, and I was informed at Jerusalem that near this a clump of native pines, *Pinus halepensis, Linn.*, exists. Maidenhair, ceterach, and the sweet Cheilanthes, were the ferns gathered, chiefly amongst the limestone clefts above Bâb el Wâd. A handsome sage, *Salvia triloba, L.*, was in flower, and several other labiates, as *Phlomis, sp.* (?);

* Professor Drummond, speaking of the slave-hunting Arabs in 'Tropical Africa' (p. 71, ed. 1888), says: 'They plant the seeds of their favourite vegetables and fruits . . . the Arab always carries seed with him . . . as if they meant to stay for ever.'

Micromeria barbata, B. et K.; M. myrtifolia, Boiss.; M. nervosa, Desf., and *Teucrium polium, Linn.*, were collected. A bryony, *B. syriaca, Boiss.*, and a beautiful clematis with whitish green flowers, *C. cirrhosa, Linn.*, trailed along the roadside walls near the villages. The leafless Ephedra and Asparagus still help to increase the variety. The spiny-branched *Calycotome villosa, Linn.*, and *Anagyris fœtida, Linn.*, yellow pea-flowered shrubs, are not uncommon. Other less important plants are : *Reseda alba, Linn.; Malcolmia crenulata, Boiss.; Thlaspi perfoliatus, Linn.; Erodium moschatum, W.; Thelygonum cynocrambe, Linn.; Ononis natrix, Linn.; Inula viscosa, Boiss.; Sherardia arvensis, Linn.; Alkanna tinctoria, Tausch.*, and *Onosma syriaca, Lab.* Most of these are common about Jerusalem and Bethlehem.

The birds noted were almost entirely British species. Of these the wheatear had not been seen before. *Saxicola lugens, Licht.*, and I think *S. Finschii, Heugl.*, were eastern chats not seen since leaving the Ghôr, but here not unfrequent.

While at Jerusalem we came in for an unusually heavy fall of snow, lasting from 20th to 25th of January. There was therefore little to be done in botany around the Holy City. Fortunately we had accomplished our pilgrimage to Jericho before the snow set in, which gave me an opportunity of comparing the northern with the southern Ghôr, or hollow of the Dead Sea.

About Jerusalem, but especially along the tiny aqueduct between the Pools of Solomon and Bethlehem, some plants were in flower. *Erodium malacoides, Linn.; E. gruinum, Linn.; Pistacia palæstina, Boiss.; Sedum sp. (S. altissimum, Poir.?); Tordylium brachycarpa, Boiss.; Torilis trichosperma, Spr.; T. leptophylla, Rich.; Pimpinella cretica, Poir.; Galium judaicum, Boiss.; Pisum fulvum, S. et L.; Lathyrus blepharicarpus, Boiss.; Carduus argentatus, Linn.; Urospermum picroides, Desf.; Crepis senecioides, Del.; Anchusa Mulleri, Willd; Onosma syriaca, Lab.; Hyoscyamus aureus, Linn.; Cyclamen latifolium, Sibth.; Plantago lagopus, Linn.; Viscum cruciatum, Linn.; Euphorbia aulacosperma, Boiss.; Gagea reticulata, R. et C.; Agrostis verticillata, Willd.*, and *Avena sterilis, Linn.;* as well as some common British plants, as *Nasturtium officinale, R. Br.; Cerastium glomeratum, Thuill.; Geranium molle, Linn.; Torilis nodosa, Gært.; Rubus discolor, W. et N.; Veronica anagallis, Linn.*, and *V*

Beccabunga, Linn., will serve to give botanists an idea of the species occurring at this season.

Jerusalem, 2,400 feet above sea-level, falls within Boissier's 'Plateaux' subdivision of the Oriental region. His 'Flora Orientalis' deals with the countries from Greece to India in a width of about twenty degrees of latitude north of the tropics; and he divides these into (1) Mediterranean, (2) Middle Europe, (3) Oriental, and (4) Region du Dattier [or Desert]. The Oriental is subdivided into Plateaux, Aralo-Caspian, and Mesopotamian. In the first of these subdivisions of the Oriental region, Jerusalem and Damascus and the districts around and above each of these cities are placed.

The climate of Jerusalem is milder and more Mediterranean than most parts of this sub-region. The date-palm, though not native nor able to ripen its fruit, can exist, and grows to goodly dimensions, as evidenced by one well-known tree. Others occur a little lower towards Ramleh. Here and at Damascus, as I subsequently saw, the prickly pear is naturalized. A 'pipi' tree, *Cæsalpinia Gilliesii*, a highland species from Buenos Ayres, was amongst the few cultivated species noticed in a recognisable condition. It was in flower beneath the windows of the Mediterranean Hotel.

From an intelligent resident at Jerusalem I obtained some information of the vegetable products of its neighbourhood which may, I think, be deemed reliable, and gives an idea of the climate.

'Frost, though occurring annually for some nights, usually at the end of January, rarely lasts throughout the day, and hardly penetrates the soil [where there is any].

'The sycamore fig, orange, mardarin orange, and lemon, which ripen their fruit so well at Jaffa, will not do so at Jerusalem.

'Apricots, tomatoes, grapes, figs (?), thrive better at Jerusalem than Jaffa. Pomegranates and nectarines do fairly well at Jerusalem.

'Bread melons [Artocarpus integrifolia, ?] and water melons, which attain a weight of 20 to 30 pounds at Jaffa, will not ripen at Jerusalem.

'A small plum, like a greengage, succeeds better at the elevated station; but strawberries, apples, and pears have all been unsuccessfully tried.

'Olives bear well about Jerusalem, especially after a winter of snow and cold; each tree generally gives a good crop every second year. Hail sometimes damages the fruit much.

'Sesame (Sesamum indicum) is grown on the plains; its oil is used for cooking purposes [and I suppose for adulterating the olive oil]. The pulp is given to animals. It is a summer crop, like the dhourra [Sorghum], after wheat and barley.'

Cupressus sempervirens, Linn., var. *pyramidalis*, the funereal cypress, attains a great size in the esplanade between the mosques of Omar and El Aksa, but far finer trees were seen later at Smyrna. The 'Prince of Wales tree,' *Pinus halepensis, Mill.*, pointed out by this name as the tree the Prince camped under, is the finest tree near Jerusalem. It is about 50 feet high, and well furnished. Smaller ones occur at the Armenian convent.

An interesting plant of Jerusalem is the red-berried mistletoe, *Viscum cruciatum, Linn.*, parasitic on olive-trees, and known elsewhere only in southern Spain. Mr. Armstrong, who was always willing (when his duties permitted) to give me a helping hand, brought me specimens from the Valley of Jehoshaphat.

During the snow at Jerusalem a gazelle was shot within a mile or two of the city. This was, I believe, a very unusual occurrence. I saw the animal immediately after its death.

CHAPTER XIII.

JERICHO AND NORTHERN GHÔR.

On January 14 we went down to the Jordan Valley. Immediately after leaving Mount Olivet I found abundance of *Androcymbium palæstinum, Baker.* (Erythrostictus, Boiss.), first seen in the 'Arabah above the Ghôr. It is a stemless white-flowered plant, small but leafy, and with rather large flowers of no particular beauty. It belongs to the Colchicaceæ. I mention it specially because Mons. Barbey states that Roth found this plant close to Jerusalem, but that after careful search he (Barbey) was unable to rediscover it. I am thus able to confirm Roth's record. Mons. Barbey's visit (April 3) was perhaps too late for the species.

On descending even a slight distance to the east the climate at once improves. Bethlehem and the neighbourhood of Solomon's Pools are distinctly milder than Jerusalem. We gradually travel from mid-winter into spring. Several plants met with before as we climbed out of the Ghôr by Wâdy Zuweirah, are again in flower as we descend. Fumaria, Carrichtera, Biscutella, Malcolmia, Erucaria, may be quoted. Fresh forms occur, as *Hypecoum procumbens, Linn.; Capsella procumbens, Linn.; Neslia paniculata, Linn.; Hippocrepis unisiliquosa, Linn.; Hymenocarpus circinnatus, Linn.; Astragalus callichrous, Boiss.; A. sanctus, Boiss.*, var.; *Trigonella arabica, Del.; Matricaria aurea, Boiss.; Chrysanthemum coronarium, Linn.; Veronica syriaca, R. et S.; Arnebia cornuta, F. et N.; Asperugo procumbens, Linn.; Emex spinosus, Camp.; Muscari racemosum, Mull.; Lamarckia aurea, Mœnch.*, and others. These are mostly small brightcoloured spring flowers. At about sea-level some desert species begin to occur, as *Zygophyllum album, Linn.* (in flower); *Prosopis Stephaniana, Willd.; Reseda pruinosa, Del.; Retama retam, Forsk.; Ochradenus baccatus, Del.; Tamarix gallica, Linn.* var., and a few more of the southern Ghôr plants.

We are again amongst the marls, and before long those of the 600 feet level, so conspicuous round the Dead Sea, can, as Professor Hull concludes, be traced, but evidently far more completely denudated in this moister and more fluviatile district. Lower marl-terraces occur, but various searches failed to bring any more sub-fossil shells to light. Canon Tristram has gathered at 250 feet in the marls near here shells identical with those obtained by us at 'Ayûn Buweirdeh.

The flora of this part of the Jordan Valley is to a certain extent a repetition of that of the southern Ghôr, but many of the interesting species are missing, and others of more familiar types take their place. Widespread European species are much more numerous. Common British species of Draba, Capsella, Thlaspi, Nasturtium, Rubus, Helosciadium, Malva, Galium, Veronica, Mentha, Solanum, Lythrum, Cichorium, Verbena, Euphorbia being all met with, in about the total of five species in the northern Ghôr to one in the southern. Nor did the season at Jericho appear to be more advanced than that at Es Safieh.

Jericho and its neighbourhood have been amply described by many able writers, and its botany has been well illustrated by Mons. Barbey in his work already referred to. This latter visitor has not, however, corrected one statement repeatedly made by various travellers, that of the ancient palm grove, extending for several miles around Jericho, there is no existing representative. There is one date-palm, 20 feet high, at Gilgal.

Of the characteristic species of the southern Ghôr growing here, I may mention *Zizyphus spina-christi, Linn.; Balanites ægyptiaca, Del.; Loranthus acaciæ, Zucc.; Calotropis procera, Willd.*, and *Populus euphratica, Oliv.*, the latter being abundant along the Jordan. This poplar is remarkable for the extraordinary variety of shapes in its leaves, especially in young trees and saplings. In full-grown trees, like the one described at the Ghôr es Safieh, they become more uniform ; ovate and slightly incised sometimes at the base, or faintly lobed in a wavy fashion. No trees were seen near Jericho in a mature condition. Tamarisk and the ' zukkum,' or false balm of Gilead (Balanites), are very abundant here. An acacia near 'Ain es Sultân was, I believe, *A. albida, Del.*, gathered previously at Gaza. It was a stunted bush, and our old friends the acacias of Sinai and Es Safieh have all disappeared except the Prosopis Stephania, a small ragged little shrub. This little ill-favoured acacia, which thrives best on saline wet places, bears a very peculiar pod, swollen, solid, and irregular, and so like a gall or deformity of some kind that it was not until opening it and obtaining its seeds I could believe it to be a natural growth.

Bananas, oranges, and a few sugar-canes are cultivated in the Arab gardens at Gilgal, the modern Jericho.

The ornithology of the Jericho district runs in parallel lines with the botany. The European sorts are much commoner than in the Ghôr es Safieh, and the tropical and Asiatic forms generally less so. Only one couple of sunbirds, and but a few of the 'hopping-thrushes' (Argya squamiceps) were seen. Shrikes were few. The palm-dove and the collared turtle were not scarce, but they were not as one to twenty here compared with those of the more southern oasis. A few bulbuls (*Pycnonotus xanthopygus, H. et Ehr.*), pied chats, *Saxicola lugens, Licht.*, and the desert blackstarts, *Cercomela melanura, Temn.*, occurred.

On the other hand, English robins, jays, chaffinches and wheatears were seen here, though not at the Ghôr es Safieh. Blackbirds, wagtails, and stonechats were commoner, and an unexpected northern visitant, a redwing, *Turdus iliacus, Linn.*, was shot at 'Ain es Sultân. This bird has not previously been obtained in Palestine, but it is likely that the wave of unusually severe weather, about to be felt by us at Jerusalem, drove many of its companions into the country.

The river Jordan was considerably swollen, and so muddy that a plunge in its waters did not look inviting. However, Laurence and I swam it and set foot on the other side of Jordan. It was about thirty yards across, with a strong current, about enough to give equal drift and headway to a swimmer. The water was too turbid for me to learn much about its inhabitants. However I picked up two molluscs, a bivalve and a univalve (*Corbicula Saulcyi, Bourg.*, and *Melanopsis costata, Oliv.*) on the muddy edge of the stream.

We returned to Jerusalem by Marsaba, where we camped on the night of the 16th—unhappily our last experience of 'tenting,' the most enjoyable kind of Eastern life. Our intended expedition by Tiberias and Merom through northern Palestine ending in Beirût was put a stop to by heavy snow. Before dismissing Jericho I have to mention the species gathered which were not previously met with: *Ranunculus asiaticus, Linn.; Matthiola oxyceras, D.C.; Saponaria vaccaria, Linn.; Silene palæstina, Boiss.; Arenaria picta, Sibth.; Rhus oxyacanthoides, Dum.; Ammi majus, Linn.; Aizoon hispanicum, Linn.; Ononis antiquorum, Linn.; Evax contracta, Boiss.; Amberboa Lippii, D.C.; Hedypnois cretica, Boiss.; Hagioseris, sp.* (?) (*H. galilæa, Boiss.* (?); *Picris, sp.: Orobanche ægyptiaca, Pers.; Linaria albifrons, Sibth.; L. micrantha, Cav.; Cuscuta, sp.* (?) (*C. palæstina, Boiss.* ?); *Convolvulus siculus, Linn.; Vitex agnus-castus, Linn.; Phalaris minor, Retz.; Schismus marginatus, P. de B.; Bromus madritensis, Linn.; Kœleria phleoides, Pers.* Of these the Orobanche was a lovely bright blue species, and the Rhus a pretty red-berried thorn very like the hawthorn, but with flattened berries and minute flowers. This thorn has been found as far south as latitude 26° in Midian, at about 4,000 feet above sea-level, by Captain Burton. The Ononis was an erect shrub, about 5 or 6 feet high, with a few slender long spiny branches and some scattered flowers like those of our own restharrow. The Ranunculus is so like Anemone coronaria (which occurred) that it was not at first distinguished from it. Both are of a gorgeous scarlet. The Vitex was one of the very few northern representatives of the tropical Verbenaceæ. It is a straggling shrub, with dull blue flowers of no beauty, and, like many other Jericho plants, found all round the Mediterranean.

Young fragments, chiefly of Cruciferæ, Leguminosæ, and Umbelliferæ, were often picked, but for these orders the season was too little advanced.

Grasses and bulbous plants were also often too young for determination.

On the way to Marsaba, a rough ride across many deep ravines, an interesting effect of aspect was noticeable. A slight greenish hue showed plainly on the hillsides with a northern aspect, while the others were as yet completely barren. In those places where the heavy dews of night are less rapidly dried up by the noonday sun, vegetation is no doubt always more abundant, the effect of shade also being to assist the early growth. An analogous effect was still more sharply defined in a different way on steep slopes looking southwards. These presented the usual monotonous barren chalky white appearance on riding upwards, where the eye only caught the outstanding bosses and prominences of rock and soil in the wâdy bed. It was difficult to recall this on looking back from above in a commanding position. The numerous little depressions and shaded hollows, with the first symptoms of incipient vegetation, gave a faint green tint to the whole. The one rested the sight, the other was a painful glare. It was about the difference between tinted and plain glass spectacles.

At Marsaba there is a date-palm tied up and supported in the courtyard of the convent, which the monks relate was planted by St. Saba (A.D. 490). Without vouching for the truth of this statement, I was interested to learn that it always bears a stoneless fruit. Of the truth of the latter information I believe there is no doubt. This convent is interesting to ornithologists as the place of the discovery of Tristram's Grakle, whose acquaintance I had first made at Mount Hor. There were several about the convent during our visit.

On the 17th we reached Jerusalem. A week later we left for Beirût, where our party divided itself, Professor Hull and his son returning homewards. Laurence and I, however, faced the snow and succeeded in crossing Lebanon and Hermon by the admirable French road to Damascus, visiting Baalbeck on the way. As I am not writing a volume of travels I will bring this part of my subject to a close. The snow lay many feet deep on these mountains reaching to Damascus and Baalbeck, so that I was unable to make any collections or observations of consequence on the natural history of this country, which is, moreover, fairly well made known by the researches of several eminent naturalists.

LIST OF SPECIES.

Missing Page

LIST OF SPECIES.

NOTE.—An asterisk (*) is placed in front of the species which do not appear to have been previously found in Palestine, about seventy in number, omitting mosses.

RANUNCULACEÆ.

Clematis cirrhosa, Linn. Jericho, Ramleh, and Jerusalem. Also at Smyrna.

Anemone coronaria, Linn. Tell Abu Hareireh and Gaza. In flower last day of December.

Ranunculus asiaticus, Linn. Jericho. Flowering in middle of January.

MENISPERMACEÆ.

Cocculus leæba, D.C., var. 'approaches C. villosus, but petals entire, fl. ♀ ' (Oliver). Trailing over acacia trees (A. tortilis) in the Ghôr el Feifeh and es Safieh, but not common.

*C. leæba, D.C., var. fl. ♂. On 'Nubk' (Zizyphus), on the east side of the 'Arabah, near El Tâbâ. This plant, with a woody stem a quarter of an inch in diameter, or more, and veined ovate-orbicular leaves, had a very distinct appearance from the foregoing wiry climber, whose leaves were smaller and narrower. I think it may be found that there are two species occurring. Cocculus leœba ranges from tropical Arabia, Nubia, and Abyssinia, to Egypt and to Sinai. An addition to the flora of Palestine. The species appears in the 'Flora Orientalis,' amongst the addenda to the fourth volume.

BERBERIDEÆ.

Leontice leontopetalum, Linn. Leaves appearing between Medjel and Jaffa in the beginning of January.

PAPAVERACEÆ.

Glaucium arabicum, Fres. Sinai, from Wâdy Lebweh to Wâdy Zelegah. Confined, so far as is known, to Sinai. Leaves only showing in November.

Hypecoum procumbens, Linn. Khân el Ahmar, between Jericho and Jerusalem.

FUMARIACEÆ.

Fumaria parviflora, Lam. Ramleh to Jerusalem ; Jericho.

F. micrantha, Lag. Wâdy Zuweirah to Bîr es Sebâ.

F. capreolata, Linn. Gaza.

CRUCIFERÆ.

Morettia canescens, Boiss. Jebel Abu Kosheibeh, near Mount Hor ; Wâdies Lebweh, Barak and es Sheikh ; El Tâbâ in the 'Arabah. Recorded from Arabia Petræa (Boiss.), and east of Gilead (Tristram).

Matthiola incana, Linn. Coast at Jaffa. A garden escape (?).

*M. *humilis*, D.C. Tell Abu Hareireh to Gaza. A rare species ; known only from Egypt, about Alexandria and Rosetta. My specimens were determined by Mons. Boissier. An addition to the flora of Palestine.

M. oxyceras, D.C. Jericho.

Eremobium lineare, Del. Wâdy 'Arabah and Wâdy Ghuweir ; Wâdy Zuweirah.

Farsetia ægyptiaca, Turr. Frequent in Sinai, and along the 'Arabah to the Dead Sea.

Nasturtium officinale, R. Br. Bethlehem and Jericho.

**Sisymbrium erysimoides*, Desf. Ghôr es Safieh and Jericho. Has been found in Sinai. A desert species, with a wide range east and west. An addition to the flora of Palestine.

Sisymbrium irio, Linn. In the same localities as the last.

Malcolmia pulchella, Boiss. From Tell Abu Hareireh westwards.

M. crenulata, Boiss. Bethlehem and Jericho; between Ramleh and Jerusalem. These two pretty little annuals appear to be confined to Syria and Palestine.

Alyssum campestre, Linn. Jericho.

A. (Koniga) libyca, Viv. Ghôr es Safieh; Bîr es Sebâ to Gaza; Ramleh to Jerusalem.

Erophila vulgaris, D.C. (Draba verna). Jerusalem to Jericho.

Notoceras canariense, R. Br. Ghôr es Safieh; Wâdy Zuweirah; Jericho.

Anastatica hierochuntina, Linn. Plentiful about 'Akabah, and above it westwards by the Haj road, and south-west to Jebel Herteh; Ghôr es Safieh.

Biscutella columnæ, Ten. Jericho; Wâdy Zuweirah to Bîr es Sebâ.

Thlaspi perfoliatum, Linn. Jericho; Bethlehem; Ramleh to Jerusalem.

Isatis (? sp.), *I. aleppica*, Scop. ? Bethlehem.

Capsella bursa-pastoris, Linn. Bîr es Sebâ, Jericho, etc.

C. procumbens, Linn. Jericho.

Erucaria aleppica, Gært. Ghôr es Safieh and north-east end of the Wâdy 'Arabah.

E. microcarpa, Boiss. Jericho; Ghôr es Safieh; plain of S. Judæa, frequent.

Neslia paniculata, Linn. Jericho.

Moricandia sinaica, Boiss. Wâdy el 'Ain in Sinai.

M. dumosa, Boiss. Wâdy Zelegah to Wâdy el 'Ain; Petra, Mount Hor, Wâdy Ghurundel, and elsewhere east of the 'Arabah. Only known previously from the Tîh.

{ *Diplotaxis harra*, Forsk. Ghôr es Safieh.
{ *D. pendula*, D.C. Petra and Ghôr es Safieh. A very variable species. This and the previous species have been thus distinguished

among my specimens by M. Boissier. The names are, I believe, synonymous; but the Petra plants had the fruits much more decidedly pendulous.

Brassica nigra, Linn. Gaza.

B. ? sp., *B. deflexa*, Boiss. ? Bethlehem. Specimens insufficient.

B. Tournefortii, Gou. Gaza. [Several indeterminable Crucifers, chiefly Brassicæ, were brought home.]

Sinapis alba, Linn. Jericho.

S. arvensis, Linn. Tell Abu Hareireh to Gaza.

Carrichtera vellæ, D.C. Wâdy Zuweirah to Bîr es Sebâ; Jericho. This species has the peculiar habit of keeping its cotyledon attached even while in flower.

*Enarthrocarpus lyratus, D.C. Bîr es Sebâ to Gaza, in several places; at Jericho. Known from Greece, Lower and Middle Egypt. An addition to the flora of Palestine.

E. ? sp. (probably *E. strangulatus*, Boiss.), but specimens imperfect. Wâdy Zuweirah to Bîr es Sebâ.

Zilla myagroides, Forsk. Frequent in Sinaitic peninsula. In flower the last week of December at the Ghôr es Safieh.

CAPPARIDEÆ.

Cleome arabica, Linn. Debbet er Ramleh, Wâdy Hamr, and Wâdy Nasb, on the west side of Sinai, and at Wâdy Zelegah, between Mount Sinai and 'Akabah.

C. trinervia, Fres. Wâdies Berrâh and Lebweh.

C. droserifolia, Del. Wâdy el 'Ain; 'Akabah, ? Ghôr es Safieh. In flower and fruit in November, as was C. arabica. C. trinervia was barely showing its leaves. 'Probably the same as C. quinquenervia, D.C.' Lowne..

Capparis spinosa, Linn., var. γ *ægyptiaca*, Boiss. Wâdies Zelegah and el 'Ain, etc. Much more glaucous and smaller (usually trailing) than the following. In Wâdy el 'Ain, where the two occur together, they look widely different. The type occurs about Jerusalem.

Capparis galeata, Fres. Cliffs above Wâdy Sarawat, and with the last. This species is sometimes an erect bush. At the base of Jebel 'Arâdah I measured one eight feet high.

RESEDACEÆ.

Ochradenus baccatus, Del. Frequent in Sinai from Wâdy Ghurundel ('Elim') to the Dead Sea.

Reseda pruinosa, Del., = *R. amblyocarpa*, Fres.? Wâdy el 'Ain to 'Akabah, and thence along the 'Arabah to the Dead Sea.

R. alba, Linn. Jericho; Ramleh to Jerusalem.

R. stenostachya, Boiss.? Wâdy el 'Ain. Too young to be certain about.

Caylusea canescens, St. Hil. Frequent in Sinai, and thence to the Dead Sea, along the 'Arabah. From the Ghôr across to Bîr es Sebâ in two or three places. An addition to the flora of Palestine.

CISTINEÆ.

Cistus villosus, Linn. Medjel to Jaffa.

Helianthemum Lippii, Pers. Frequent in the upper parts of Sinai, and along the edge of the Tîh to 'Akabah; Mount Hor; Gaza. A variety, with larger pedunculate flowers, occurs at Jebel Abu Kosheibeh. A late summer flowerer, almost all dead in December.

H. kahiricum, Del. Wâdy Zuweirah to Bîr es Sebâ. A very early flowerer, just appearing at the end of December.

SILENEÆ.

Dianthus multipunctatus, Ser. Wâdy Ghurundel (Edom) and Mount Hor; plain of Judæa from Wâdy Zuweirah to Gaza. Not found so far south previously.

D. sinaicus, Boiss. Mountains about Wâdy es Sheikh. Confined to Sinai.

Saponaria vaccaria, Linn. Jericho.

Gypsophila rokejeka, Del. With Dianthus sinaicus at Zibb el Baheir in Wâdy Lebweh; Wâdy Berrâh; Jebel Abu Kosheibeh; 'Ayûn Buweirdeh in the 'Arabah.

G. (Saponaria) hirsuta, Lab., δ *alpina* (Boiss.). Summit of Jebel Katharîna at 8,500 feet. One of the very few species of the higher Mediterranean alps found in Sinai. This particular form is confined to Lebanon, Hermon, Makmel, and the present station.

Silene dichotoma, Ehrh. Wâdy Zuweirah to Bîr es Sebâ.

S. atocion, Murr., *S. ægyptiaca*, Linn. Ramleh to Jerusalem; Jericho.

*S. Hussoni, Boiss. Between Bîr es Sebâ and Tell Abu Hareireh. Coming into flower at the end of December. Mons. Boissier, who named this species for me, records it from a single locality, 'Ouadi Sannour deserti Ægyptiaco-Arabici.' An addition to the flora of Palestine.

S. inflata, Sm. Gaza.

S. palæstina, Boiss. Jericho.

*S. colorata, Poir. Jericho. Not included in the 'Flora Orientalis,' but found in Crete, according to Nyman. An addition to the flora of Palestine.

S. succulenta, Forsk. Sands at Gaza.

Buffonia multiceps, Dcne. Near the summits of Jebel Katharîna and Jebel Mûsa; Zibb el Baheir at the head of Wâdy Lebweh. Confined to Sinai.

Arenaria (Alsine) picta, Sibth. Jericho.

A. graveolens, Schreb. Summits of Jebel Mûsa and Jebel Katharîna; top of Zibb el Baheir, in Wâdy Lebweh.

Stellaria media, Linn. Gaza.

Cerastium glomeratum, Thuill. Bethlehem.

Spergula pentandra, Linn. Jericho and Ghôr es Safieh. Noticed previously at Gaza.

Spergularia marginata (Koch.). Saline swamps at Tell el Milh (Moladah).

PARONYCHIACEÆ.

Polycarpæa fragilis, Del. Frequent in the lower parts of Sinai, and along the 'Arabah Valley to the Ghôr es Safieh.

P. (Robbairea) prostrata, Dcne. Same range as last, but less common. At Jebel Usdum. Plentiful at Debbet er Ramleh, in Sinai.

**Polycarpon succulentum*, Del. At Jericho, and between Tell Abu Hareireh and Gaza. An addition to the Palestine flora.

Herniaria hemistemon, J. Gay. (?). Jericho. Too young to be certain about.

*Paronychia nivea, D.C., = *P. capitata*, Lam., not Koch. About Bîr es Sebá. Not included in the 'Flora Orientalis,' but found as far east as Greece, according to Nyman. An addition to the flora of Palestine.

P. argentea, Lam. Mount Hor. A variety of this species occurred at a considerable elevation. Gaza; Jericho.

**P. desertorum*, Boiss. 'Ayûn Mûsa, near Suez; Wâdy Ghuweir, on the east or Edomitic side of the 'Arabah. Perhaps only a form of P. arabica, Linn. An addition to the flora of Palestine.

Gymnocarpum fruticosum, Linn. Common in Sinai and along the 'Arabah, especially in the Edomitic valleys.

**Sclerocephalus arabicus*, Boiss. Ghôr es Safieh. A desert species found in Sinai. An addition to the flora of Palestine.

Pteranthus echinatus, Desf. Jericho.

CERATOPHYLLEÆ.

Ceratophyllum demersum, Linn. Drifting in the Gulf of Suez at 'Ayûn Mûsa.

MOLLUGINEÆ.

Glinus lotoides, Linn. Wâdies Nasb and Sarawat. Not noticed hitherto in Sinai.

TAMARISCINEÆ.

Tamarix gallica, Linn., var. *nilotica*, Ehr., et *T. nilotica*, Ehrh. Frequent in Sinai to the Dead Sea, and in the Jordan Valley. Var.

mannifera, Ehr., occurs also, as on the summit of Jebel Usdum and Wâdy Zelegah.

*Tamarix articulata, Vahl. Between Ramleh and Jerusalem; ? introduced. A handsome tree about Gaza, thirty-five feet high, belongs, I believe, to this species. An addition to the flora of Palestine.

Reaumuria palæstina, Boiss. Common in Sinai. This form has been recorded only from the Dead Sea, but probably, as suggested by Lowne, it is identical with *R. hirtella*, Jaub., *R. vermiculata*, Linn., the name by which the Sinai plant has been collected. Flowers were obtained in November.

HYPERICINEÆ.

Hypericum tetrapterum, Fr. *forma*. Between Gaza and Tell Abu Hareireh. A small prostrate form, but not in flower.

MALVACEÆ.

Malva rotundifolia, Linn. Wâdies Nasb and Sarawat; Jebel Mûsa; Petra; very abundant about Tell el Milh and Bîr es Sebâ. Eaten by the Bedawin.

M. sylvestris, Linn. Jericho.

Abutilon muticum, Del. Ghôr el Feifeh and Ghôr es Safieh.

A. fruticosum, G. et P. (*Sida denticulata*, Fres.). Wâdy el 'Ain in Sinai, and Wâdy Ghurundel in Edom.

TILIACEÆ.

Corchorus trilocularis, Linn. Ghôr es Safieh and Feifeh, frequent.

GERANIACEÆ.

Geranium tuberosum, Linn. I gathered the tuberous roots of, I believe, this species, like small flattened potatoes, on the summit of Mount Hor.

G. molle, Linn. Bethlehem.

Erodium cicutarium, Linn. From the Ghôr across Judæa to Gaza, and at Jericho.

E. moschatum, Linn. Jericho; Ramleh to Jerusalem.

Erodium gruinum, Boiss. Bethlehem.
E. laciniatum, Cav. Jericho.
E. malacoides, Linn. Jericho and Bethlehem.
E. hirtum, Forsk. Frequent in Sinai; in the 'Arabah to the Ghôr, and in the Edomitic valleys; Tell Abu Hareireh to Gaza.
E. glaucophyllum, Ait. Wâdies Sudur and Ghurundel (Elim), on the north-west side of Sinai.
E. (?) sp. *bryoniæfolium*, Boiss.? Medjel to Jaffa.
Monsonia nivea, Dcne. Debbet er Ramleh; Wâdies Lebweh and Berak; 'Ayûn Buweirdeh and near the Ghôr in the 'Arabah.

ZYGOPHYLLEÆ.

Tribulus terrestris, Linn. Wâdies Lebweh and Berak; W. Ghuweir in Edom; Ghôr.
T. alatus, D.C. Wâdy el 'Atttyeh and Jebel Herteh.
Fagonia glutinosa, Del. Debbet er Ramleh; Wâdies Ghurundel (Elim) and Sudur.
F. myriacantha, Boiss. Wâdies Lebweh, es Sheikh, Harûn, and 'Arabah.
F. kahirica, Boiss. Ghôr es Safieh.
F. grandiflora, Boiss. Wâdy 'Arabah, near the Ghôr.
F. arabica, Linn. Debbet er Ramleh to 'Akabah, common in Sinai.
F. cretica, Linn. var. Ghôr es Safieh and Wâdy Ghuweir.
F. cretica, Linn., 'var. ramulis subteritis incano-puberalis,' Oliver. Wâdy el 'Ain and Jebel Usdum. As deserving of special distinction as several of the above. Fresh varieties turn up in every wâdy, and it is hopeless to endeavour to separate the numerous described varieties. It would perhaps be preferable to follow Anderson in re-uniting them under the general name. The above are Mons. Boissier's determinations of my specimens.

Zygophyllum dumosum, Boiss. Wâdy Hessi and Jebel Herteh to 'Akabah.

Zygophyllum simplex, Linn. Ghôr es Safieh. An addition to the flora of Palestine.

Z. album, Linn. Wâdy Ghurundel (Elim) and elsewhere; very frequent in Sinai and about the Dead Sea. Between Jerusalem and Jericho at about sea-level.

Z. coccineum, Linn. Wâdy el 'Ain.

Seetzenia orientalis, Dcne. Debbet er Ramleh and Wâdy Hamr.

Peganum harmala, Linn. Wâdies el 'Ain, es Sheikh, Berrâh, and Lebweh; summit of Jebel Mûsa and Ghôr es Safieh.

Nitraria tridentata, Desf. 'Ain Mûsa, Wâdy Ghurundel (Elim), Wâdy es Sheikh, 'Akabah, and by the Dead Sea.

RUTACEÆ.

Ruta graveolens, Linn. (*R. chalepense*, Linn.). Jericho; Medjel to Jaffa.

R. (*Haplophyllum*) *tuberculatum*, Forsk. Wâdy Zelegah to Wâdy el 'Ain; 'Akabah; frequent in Edomitic wâdies; Ghôr es Safieh. A hairy form occurs at Wâdy Lebweh and Debbet er Ramleh. This plant has a most sickening and persistent smell.

SIMARUBEÆ.

Balanites ægyptiaca, Del. Jericho, and Ghôr es Safieh and Feifeh.

TEREBINTHACEÆ.

Rhus oxyacanthoides, Dum. Several bushes between Jericho and the Jordan, the ''Arûr' of Bedawîn. Has much the appearance of our hawthorn, but with inconspicuous flowers, and the red berries flattened.

Pistacia palæstina, Boiss. Bethlehem; summit of Mount Hor, at an estimated height of 4,400 feet. Not found so far south previously.

Zizyphus spina-christi, Linn. Wâdy Nasb; 'Akabah; Wâdy Ghuweir. Abundant about Jericho and in Ghôr el Feifeh, and Ghôr es Safieh.

Rhamnus punctata, Boiss., ? var. *microphylla*, Boiss. Medjel to Jaffa. Barren.

Rhamnus punctata, Boiss., and *R. palæstina*, Boiss. ? Medjel to Jaffa.

R. sp., *R. oleoides*, Linn. ? In several places in a barren condition in the Ghôr es Safieh. Leaves small and deciduous. Shrub six to ten feet high, and called by the Bedawin ' Seisaban.'

LEGUMINOSÆ.

Anagyris fœtida, Linn. Gaza ; Ramleh to Jerusalem.

Crotalaria ægyptiaca, Bth. Wâdies Nasb and Sarawat.

Lupinus reticulatus, Desv. Between Ramleh and Jerusalem, and at Ascalon, the eastern limits of the species.

L. termis, Forsk. About Medjel, Gaza and Jaffa. Cultivated.

Lotononis lebordea, Benth. (*L. dichotoma*, Del.) Wâdies Lebweh, Berrâh, es Sheikh, and 'Arabah.

Calycotome villosa, Linn. Sand hills at Ascalon ; Medjel ; Ramleh to Jerusalem.

Genista (*Retama*) *retam*, Forsk. Common throughout Sinai and South Palestine. The juniper (' rothem ') of the Bible.

Ononis antiquorum, Linn. 'Ain es Sultân, Jericho. An erect shrub, with long stiff spiny branches, bearing scattered sessile flowers, four to six feet high.

*O. *campestris*, Koch. (*O. spinosa*, Linn.) Gaza, prostrate or decumbent on the sands. Perhaps not specifically distinct from the last, but widely so in habit. An addition to the flora of Palestine.

O. natrix, Linn., var. γ *stenophylla*, Boiss. Gaza ; Ramleh to Jerusalem.

O. vaginalis, Vahl. From Petra to about 4,000 feet on Mount Hor. Found hitherto only at Alexandria and in plains under Anti-Lebanon in the Oriental region.

O. serrata, Forsk., var. *major* ? Bîr es Sebâ to Gaza in several places.

Trigonella arabica, Del. Jericho.

Medicago denticulata, Willd. Ghôr es Safieh ; Wâdy Ghuweir.

Medicago laciniata, All. Between Bîr es Sebâ and Tell Abu Hareireh ; Gaza. This minute trefoil was in flower at the end of December.

Hymenocarpus circinnatus, Linn. Jericho.

Lotus tenuifolius, Rchb., γ *uniflorus*, Boiss. (*L. corniculatus*, var. *L.*) Swamps in the Ghôr es Safieh, near the Dead Sea.

L. lanuginosus, Linn. Wâdy el 'Attîyeh and Jebel Herteh.

L. lamprocarpus, Boiss. Ghôr es Safieh and Jericho.

Hippocrepis unisiliquosa, Linn. var. Jericho.

Psoralea bituminosa, Linn. Ghôr es Safieh.

P., ? sp., *P. plicata*, Del. ? Zibb el Baheir and Jebel Watîyeh ; Jebel Abu Kosheibeh.

Indigofera argentea, Linn. Ghôr es Safieh and el Feifeh. Cultivated, but now spontaneous.

**I. paucifolia*, Del. Ghôr es Safieh. Upper Egypt is its nearest locality in Boissier's 'Flora Orientalis,' but Oliver enumerates it amongst Captain Burton's 'Indian Plants.' An addition to the flora of Palestine.

Tephrosia, ? sp., *T. purpurea*, Pers. ? Wâdy el 'Attîyeh and Jebel Herteh to 'Akabah. Apparently this species, but not in good condition. Not mentioned in Boissier's 'Flora Orientalis.'

T. apollinea, Del. 'Akabah ; Wâdy el 'Ain.

*Colutea aleppica, Lam. Summit of Mount Hor. An addition to the flora of Palestine. Probably this species, but specimens imperfect, owing to season.

Astragalus macrocarpus, D.C. ? A little south of Gaza. Not in fruit.

A. callichrous, Boiss. Jericho.

A. alexandrinus, Boiss. Wâdy Zuweirah to Bîr es Sebâ.

**A. acinaciferus*, Boiss. On the marl banks by watercourses immediately above the Ghôr in the 'Arabah. An addition to the flora of Palestine.

Astragalus sieberi, D.C.   Debbet er Ramleh ; Wâdies Hamr, Berrâh, and Lebweh.

A. trigonus, D.C. ?   Wâdies Berrâh and Lebweh.

A. aleppicus, Boiss.   Gaza.   Not recorded so far south previously.

A. Forskahlii, Boiss.   Wâdy 'Arabah and Ghôr es Safieh.

A. sanctus, Boiss.   Wâdy Zuweirah to Bîr es Sebâ.

A. sanctus, Boiss., var. *stenophylla* ?   Jericho.

Onobrychus ptolemaica, Del.   Wâdy Lebweh ; 'Akabah.

Alhagi maurorum, D.C.   'Ayûn Mûsa ; Wâdies el 'Ain, es Sheikh, and Ghôr es Safieh.

Vicia sativa, Linn., *forma*.   About the Dead Sea.

V. palæstina, Boiss.   With the last.

Lathyrus blepharicarpus, Boiss.   Bethlehem.

Pisum fulvum, S. et S.   Gaza and Bethlehem.

**Rhynchosia minima*, D.C.   Ghôr es Safieh, where it covered acacia trees in one or two places by the river with canopies of pretty little flowers.   Not included in Boissier's ' Flora Orientalis,' nor found elsewhere so far north of the tropics.   ' A common tropical plant in both hemispheres,' Hooker, Fl. Nigrit.   R. schimperi has been lately found in Midian by Burton.

Dolichos lablab, Linn.   Ghôr es Safieh in several places, especially along the stream.   Widely diffused in warm countries, where it is commonly eaten.

Cassia obovata, Collad.   'Akabah.   [C. bicapsularis, Linn.   Cultivated at 'Ayûn Mûsa.]

C. acutifolia, Del.   (*C. lanceolata*, Forsk.)   'Akabah.   Kosseir and Assouan in Upper Egypt are its nearest known localities.   An addition to the flora of Sinai.

Ceratonia siliqua, Linn.   First seen at Gaza ; frequent northwards.   The Caroub, or St. John's locust tree.

Prosopis stephaniana, Willd.   Wâdy Ghuweir, Ghôr es Safieh, Gaza, and Jericho.   Very abundant near the Dead Sea, and bearing its monstrous-looking pods in January.

Acacia tortilis, Hayne. Debbet er Ramleh ; Wâdy Nasb ; 'Akabah ; east side of the 'Arabah at El Tâbâ. Frequent in the Ghôr, where, though only ten or fifteen feet high, it has a thick trunk, and gives considerable shade.

A. seyal, Del. Frequent in Sinai. Commoner than the last on the west side, in the hotter sandy and arid places. Smaller and more fiercely spiny than A. tortilis, and seldom more than six or seven feet high ; usually less, with a flat tabulated top. Leaves more finely pinnated, and bark less red than in the last. Pods different.

A. albida, Del. Sands at Gaza, close to the Quarantine ground ; also, I think, this species at 'Ain es Sultân, Jericho. Perhaps not native. Inhabits tropical Africa.

**A. læta*, Br. Sparingly in the Ghôr es Safieh. A handsome tree, larger than the other species met with, 20 to 25 feet high. Leaflets considerably larger and fewer, and pods very different from A. tortilis or 'seyal.' An Arab of the Ghôr whom I questioned recognised it as distinct, calling it ' sunt,' and pointed out its difference in leaves and pod from the 'seyal.' There have been very confusing statements made about the names, which arise from the Bedawin themselves. A Bedawin at 'Akabah called a remarkably fine specimen of A. tortilis by the Egyptian term, ' sont.' A. læta has not been found hitherto nearer than Assouan, seven degrees south, and is an important addition to the flora of Palestine.

[N.B.—Many fragments of Medicago, etc., not yet in flower. Trefoils had hardly appeared.]

ROSACEÆ.

Cratægus sinaica, Boiss. Near the summit of Jebel Mûsa.

C., ? sp., *C. azarolus*, Linn. ? Near Gaza.

Rubus discolor, W. et N. (*R. sanctus*, Schreb.) Bethlehem and Jericho.

Poterium verrucosum, Ehr. ? Wâdy Ghuweir, Gaza. Not in flower.

P. spinosum, Linn. Judæa and northwards, common.

Neurada procumbens, Linn. Wâdy Lebweh, Gaza.

FAUNA AND FLORA OF SINAI, PETRA, AND WADY 'ARABAH. 93

LYTHRARIEÆ.

Lythrum salicaria, Linn., var. *tomentosum*, D.C. Twelve to fourteen feet high in the Ghôr es Safieh.

L. hyssopifolium, Linn. Ghôr es Safieh and Jericho, in wet places.

[*Lawsonia alba*, Linn. Hedge-rows near Gaza. Cultivated at 'Akabah. Not native?]

CUCURBITACEÆ.

Cucumis prophetarum, Linn. Frequent in Sinai to 'Akabah, and thence to the Dead Sea.

C. trigonus, Roxb. Ghôr es Safieh. Several of this family cultivated in the Ghôr.

Citrullus colocynthis, Schr. Sinai, and thence to Ghôr; near Bîr es Sebâ.

Bryonia syriaca, Boiss. Between Ramleh and Jerusalem. A bryony occurred on the summit of Mount Hor; I believe, of this species; but it was not in flower.

B. multiflora, Boiss. Gaza.

FICOIDEÆ.

Aizoon hispanicum, Linn. Jericho.

A. canariense, Linn. Ghôr es Safieh; Wâdy Mûsa.

*Trianthema pentandra, Linn. Ghôr es Safieh. A native of Senegal, Nubia and tropical Arabia, and found also in Afghanistan and Persia at Bushire. An addition to the flora of Palestine. Has been found in Midian by Captain Burton.

CRASSULACEÆ.

Cotyledon umbilicus, Linn. ? *Umbilicus intermedius*, Boiss. ? Jebel Watîyeh and Zibb el Baheir, west of Mount Sinai. Plentiful on Mount Hor, and on Jebel Abu Kosheibeh. Apparently the same species was common about Bethlehem, but it was nowhere in determinable condition.

The root-stock was tuberous. Its leaves did not appear to differ from those of our British species. Recorded by Decaisne under that name.

Ledum, ? sp., *L. altissimum*, Poir. ? Solomon's Pool, near Bethlehem. Too young to name.

UMBELLIFERÆ.

Eryngium, ? sp. Wâdy Ghurundel, east side of Wâdy 'Arabah. Too young to determine.

Helosciadium nodiflorum, Linn. Near Jericho.

Bupleurum linearifolium, D.C. Jebel Mûsa; apparently this species also near Bîr es Sebá.

Deverra tortuosa, Desf. Frequent in Sinai, extending up the 'Arabah to the Ghôr, and across Judæa to Gaza.

Pimpinella cretica, Poir. Bethlehem.

Torilis nodosa, Gært. Bethlehem.

Chætosciadium trichospermum, Linn. Bethlehem; Jericho.

Caucalis leptophylla, Linn. Bethlehem.

Ammi majus, Linn. Jericho.

Carum, ? sp. Deir el 'Arbain, and near the convent on Jebel Katharína.

Ainsworthia trachycarpa, Boiss.

RUBIACEÆ.

*Rubia peregrina, Linn. ? Petra and stony slopes of Mount Hor, to Wâdy Mûsa. The specimens were barren, but I thought unmistakable. Unfortunately in the hurry of my visit to Petra, I omitted to preserve specimens of this plant. An addition to the flora of Palestine.

R. Olivieri, A. Rich. Ramleh to Jerusalem; Gaza.

Sherardia arvensis, Linn. Ramleh to Jerusalem.

Galium sinaicum, Dcne. Wâdy es Sheikh and Jebel Mûsa. Confined to Sinai.

G. canum, Req. Petra and Mount Hor. Not found hitherto so far south.

PALESTINE EXPLORATION FUND Pl.1

W H Fitch del. et lith. Hanhart imp

GALIUM PETRÆ, n.sp. 2, DAPHNE LINEARIFOLIA.

Galium petræ, sp. nov. hispidulum, caulibus diffusis elongatis 4-angulatis fragilibus plus minusve intricatis, foliis quaternis angusti linearibus uninerviis rigidiusculis, cymis sæpius paucifloris axillaribus terminalibusque, folio 2–4 plo longioribus pedicellis gracilibus divaricatis hispidulis, fructu (immaturo) pilis brevibus incurvis uncinatisve hispido. Hab. in Wâdy Mûsa, apud Petram. Foliis 1½–2 lineæ longis setiformibus, internodis inferioribus 7–8 lineæ longis, verticillatis foliorum supremis approximatis. 'Affinis G. jungermannioides sed foliis quaternis nec senis, caule ramisque elongatis nec pulvinatis . . . . .' E. Boissier. [Plate XVI., fig. 1.]

G. aparine, Linn. Jericho and Gaza.

G. judaicum, Dcne. Bethlehem. Confined to Palestine, from Lebanon to Jerusalem.

DIPSACEÆ.

Pterocephalus sanctus, Dcne. Jebel Mûsa; Deir el 'Arbain and Jebel Katharina, Petra, and Mount Hor. An addition to the flora of Palestine. Hitherto found only on Sinai.

COMPOSITÆ.

Bellis sylvestris, Cyr. Rocky places about Bâb el Wâd, between Ramleh and Jerusalem.

Asteriscus pygmæus, Coss. Abundant on the Edomite escarpment, at the head of Wâdy Ghurundel; Ghôr es Safieh.

A. graveolens, Forsk. Frequent on Sinai.

Anvillæa garcini, D.C. Wâdy Harûn.

Inula crithmoides, Linn. Ghôr es Safieh, 4 to 7 feet high.

I. viscosa, Desf. Wâdies east of the 'Arabah; Ghôr es Safieh; between Ramleh and Jerusalem.

I. dysenterica, Linn. Ghôr es Safieh.

Pulicaria arabica, Cass. Ghôr es Safieh; 'Ayûn Buweirdeh in the 'Arabah.

P. (Francoeuria) crispa, Forsk. Jebel Mûsa; Wâdy el 'Ain; Ghôr es Safieh.

Pulicaria undulata, D.C. Throughout Sinai to the Dead Sea.

Iphiona juniperifolia, Cass. Frequent at about the mean elevation of Sinai, as at Wâdy es Sheikh, etc.

I. scabra, D.C. Wâdies Zelegah and el 'Ain.

**Varthamia montana*, Vahl. Wâdies Lebweh, Berrâh, and es Sheikh; Mount Hor. An addition to the flora of Palestine. Known previously to my visit from Sinai only.

Pluchea dioscorides, D.C. Wâdy Ghuweir; Ghôr es Safieh. A bushy composite, 12 to 15 feet high, in Wâdy Ghuweir.

**Erigeron (Conyza) bovei*, D.C. Wâdy el 'Ain and Ghôr es Safieh. Attaining the height of 10 feet in Wâdy el 'Ain amongst tamarisk bushes. An addition to the flora of Palestine.

Phagnalon nitidum, Fres. Wâdies Lebweh and es Sheikh.

Leyssera capillifolia, Willd. Frequent on Sinai and along the 'Arabah to the Dead Sea.

Evax contracta, Boiss. Jericho.

E. anatolica, Boiss. Jebel Usdum.

Filago prostrata, Parlat. Wâdy Hessi, south-west of 'Akabah.

Ifilago spicata, Forsk. Along the Tîh escarpment, in many places between Sinai and 'Akabah; Wâdy Harûn.

*Eclipta alba, Linn. Ghôr es Safieh. An addition to the flora of Palestine. A tropical and sub-tropical species found in Egypt.

Xanthium strumarium, Linn. ? Gaza.

Achillea santolina, Linn. Wâdy Zuweirah to Bîr es Sebâ.

Anthemis, ? sp. Unrecognisable young plants of this genus, and the last, occurred at Jericho and elsewhere.

Matricaria aurea, Boiss. Jericho.

M. auriculatum, Boiss. Ghôr es Safieh.

Chrysanthemum coronarium, Linn. Jericho.

Pyrethrum santalinoides, D.C. Wâdy es Sheikh; Jebel Mûsa.

Cotula cinerea, Del. Wâdy Ghurundel ('Elim').

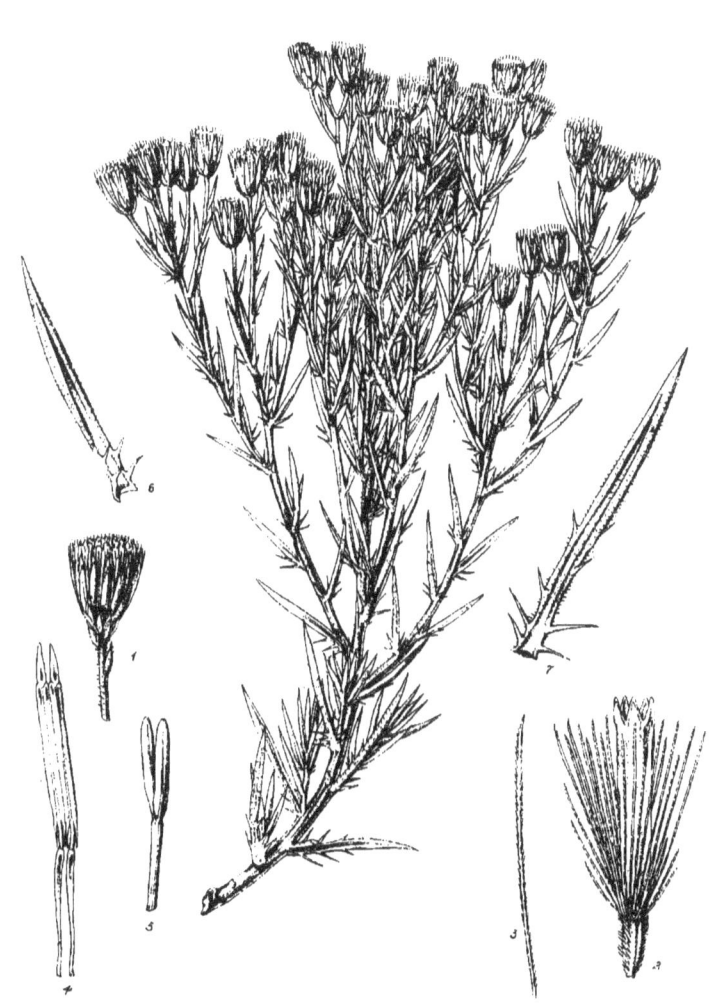

IPHIONE SCABRA. D.C

Artemisia monosperma, Del. Abundant at 'Akabah; 'Ayûn Buweirdeh in the 'Arabah; Gaza.

A. herba-alba, Asso., var. *laxiflora*. Wâdy es Sheikh.

A. herba-alba, Asso. Abundant with several varieties from Wâdy Lebweh to Wâdies Zelegah and el 'Ain, and occurs between Bîr es Sebâ and Tell Abu Hareireh. At Wâdy el 'Ain a variety, with densely aggregated reddening flower-heads, occurs.

A. judaica, Linn. Wâdy Sudur. Several indeterminable Artemisiæ were gathered.

Santolina fragrantissima, Forsk. Abundant at a moderate elevation in Sinai, but very scarce after 'Akabah.

Senecio vernalis, W.K. Gaza and Jericho.

S. coronopifolius, Desf. Ghôr es Safieh, and across Judæa to Gaza.

Calendula arvensis, Linn. Wâdy Zuweirah to Gaza.

**Tripteris Vaillantii*, Dcne. Petra and lower slopes of Mount Hor. An addition to the flora of Palestine.

**Echinops glaberrimus*, D.C. Wâdies Lebweh, es Sheikh, and Harûn; Zibb el Baheir. Wâdy Harûn brings the range within Canon Tristram's limits allotted to Palestine, to which flora it is an addition. Hitherto believed peculiar to Sinai.

Carduus argentatus, Linn. Bethlehem.

Atractylis flava, Desf. Wâdy Hessi, a little south of the Haj road, near 'Akabah; at 'Ain Mûsa.

A. prolifera, Boiss. Gaza.

Amberboa Lippii, D.C. Jericho; noted from the Tîh desert, but not from Palestine Proper previously.

A. crupinoides, Desf. Ghôr es Safieh; Jericho.

Centaurea sinaica, D.C. Wâdy Hessi; apparently this species, but withered.

C. araneosa, Boiss. Gaza; Medjel to Jaffa.

C. iberica, Trev. Ghôr es Safieh.

C. pallescens, Del. Gaza.

Centaurea eryngoides, Lam. Zibb el Baheir; Jebel Mûsa; Mount Hor.

C. (Phæopappus) scoparia, Sieb. Wâdies Lebweh and es Sheikh; Jebel Mûsa.

Carthamus lanatus, Linn. Wâdy Harûn.

C. glaucus, M.B. Wâdy 'Arabah; Bîr es Sebâ to Tell Abu Hareireh.

Cichorium intybus, Linn. Jericho. In flower in the middle of January.

Hedypnois cretica, Linn. Jericho.

Tolpis altissima, Pers. Gaza and Tell Abu Hareireh.

Thrincia tuberosa, Linn. With the last. At a little distance closely resembles our dandelion.

Hagioseris, ? sp. *H. galilæa*, Boiss. ? Jericho. *Picris*, ? sp. Jericho.

Urospermum picroides, Linn. Bethlehem.

**Scorzonera alexandrina*, Boiss. Tell Abu Hareireh to Gaza. Specimens determined by Mons. Boissier. Hitherto found only at Tunis, Algiers, and Alexandria. An addition to the flora of Palestine.

S. lanata, M.B. With the last species, and about Bîr es Sebâ.

**Sonchus maritimus*, Linn. Sparingly near the Dead Sea in the Ghôr es Safieh. An addition to the flora of Palestine.

Zollikoferia, ? sp., *Z. casiniana*, Jaub. ? Apparently this species, from Mount Hor and Ghôr es Safieh. Recorded from Egypt and the shores of the Red Sea at Kossni. An addition (if correct) to the flora of Palestine.

Z. stenocephala, Boiss., vel *Z. arabica*, Boiss. Specimens incomplete. Wâdy Zuweirah. The first-mentioned form has been collected only in Beloochistan.

Z. nudicaulis, Linn. 'Akabah; Wâdy el 'Ain; Ghôr es Safieh.

Z. spinosa, Forsk. Wâdies Nasb and Lebweh; 'Akabah.

**Crepis senecioides*, Del. Bethlehem, near Solomon's Pool. The localities known to Boissier are Alexandria, Cairo, and Sackara in Egypt. An addition to the flora of Palestine.

ERICACEÆ.

Arbutus andrachne, Linn. Limestone cliffs above Bâb el Wâd, between Ramleh and Jerusalem.

PRIMULACEÆ.

Anagallis arvensis, Linn. Jericho.
A. arvensis, Linn., var. *cærulea*. Jericho; Bethlehem.
A. latifolia, Linn., = var. *A. arvensis* ? Ghôr es Safieh.
Lysimachia dubia, Ait. Ghôr es Safieh, near the Dead Sea.
Cyclamen latifolium, Sibth. First seen near Bîr es Sebâ; thence to Jerusalem. Common.
Primula boveana, Dcne. Wâdy es Sheikh. Not in flower.

OLEACEÆ.

Olea europea, Linn. Olive-trees were measured a little north of Gaza, having a circumference of 18 and 20 feet a couple of feet above the ground. These were the largest and oldest trees seen. Schweinfurth has suggested that this tree, perhaps the oldest cultivated, was originally derived from the Nubian mountains, where he has found it native.

SALVADORACEÆ.

Salvadora persica, Garcin. Ghôr el Feifeh and es Safieh. Very abundant, forming dense groves, and giving off a heavy smell, resembling that of a monkey-house. Found also in Sinai.

APOCYNEÆ.

Nerium oleander, Linn. Wâdies Mûsa, Harûn, Ghuweir, and Ghurundel, in Edom; and in the wâdies of Moab leading to the Ghôr el Feifeh and es Safieh; Jericho.

ASCLEPIADEÆ.

Calotropis procera, Willd. Ghôr el Feifeh and Safieh. Found also in Sinai.

Pentatropus spiralis, Forsk. Ghôr es Safieh and Wâdy Harûn. Muscat, in South-east Arabia, is the only habitat in the region in Boissier's 'Flora Orientalis.' A native of tropical countries and an addition to the flora of Palestine.

Dæmia cordata, Br. Frequent in Sinai and along the 'Arabah to the Dead Sea. In most of the Edomitic wâdies and about 'Akabah, as also in the wâdies on the west side of the 'Arabah. In Wâdy Harûn climbing to 10 or 12 feet in height on Ochradenus.

Gomphocarpus sinaicus, Boiss. Wâdies Lebweh and es Sheikh; 'Ain Abu Zuweirah and Mount Sinai, near the convent; 'Akabah; Jebel Abu Kosheibeh. In cultivation in a garden at Jaffa.

Cynanchum acutum, Linn. Climbing over reeds at 'Ayûn Buweirdeh in Wâdy 'Arabah; Wâdy Ghuweir; Ghôr es Safieh; by the Jordan, at the pilgrims' bathing-place. Only mentioned from the maritime districts of Palestine by Tristram.

Leptadenia pyrotechnica, Forsk. Ghôr el Feifeh and es Safieh sparingly; by the Dead Sea, at the northern end of Jebel Usdum.

**Boucerosia aaronis*, sp. nov. Herba succulenta, 4–6 pollicaris, ramis erectis parce ramosis rigidis 4-gonis angulis acutis remote tuberculato dentatis, floribus livido-purpureis, corolla notata, limbo profunde 5-fido lobis ovato-deltoideis intus glabris vel obsolete puberulis, coronæ stamineæ lobis interioribus oblongis obtusissimis incumbentibus, exterioribus bifidis dentibus divaricatis, lobis interioribus paullo brevioribus. Fl. Dec.

Hab. In aridis saxosis locis, Montium Aaronis et 'Abu Kosheibeh.'

Tota planta fere pallida-fusca, carnosæ, aphylla, facie *Stapeliæ*. (Descriptiones specierum Boucerosiæ in 'Flora Orientale' minime sufficientes sunt.) Hæc species ab eis videtur distincta. [Plate XVII., figs. 1 to 8.]

GENTIANACEÆ.

*Erythræa spicata, Pers. Ghôr es Safieh. An addition to the flora of Palestine.

CONVOLVULACEÆ.

Convolvulus lanatus, Vahl. Wâdy Nasb.

C. siculus, Linn. Jericho to Marsaba.

GOMPHOCARPUS SINAICUS, Boiss

PALESTINE EXPLORATION FUND PL 4

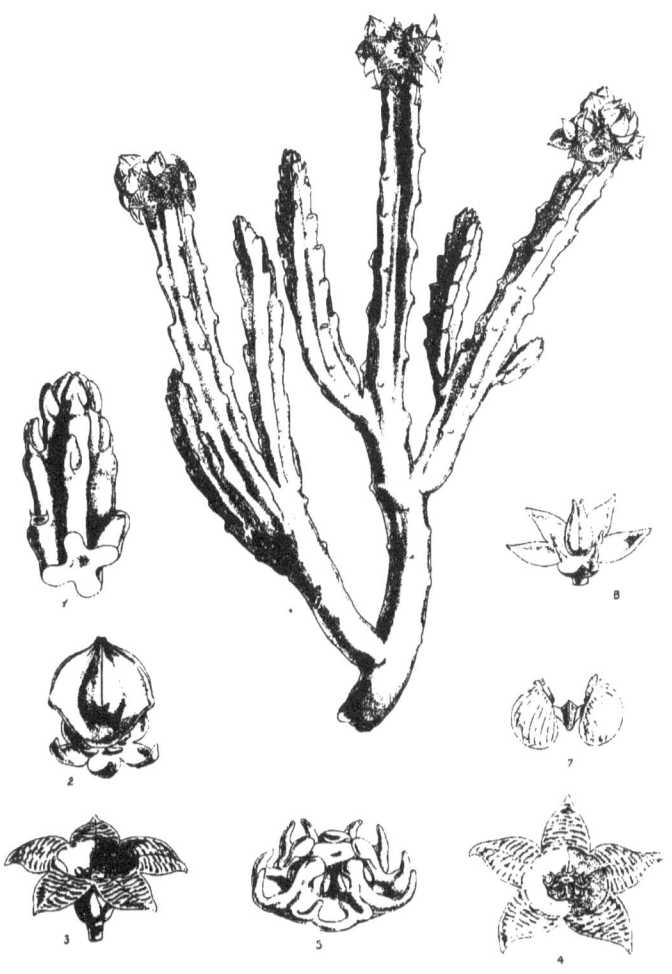

W.H.Fitch del
C.H.Fitch lith

BOUCEROSIA AARONIS, HART

Hanhart imp

Convolvulus arvensis, Linn. Gaza and Jericho. Several other indeterminable convolvuli from Sinai and Gaza.

Cressa cretica, Linn. 'Ain Mûsa ; Wâdy Ghurundel (Elim) ; 'Akabah ; 'Ayûn Buweirdeh in the 'Arabah ; Ghôr es Safieh.

Cuscuta, sp., ? *C. palæstina*, Boiss. On mint at Jericho.

BORAGINEÆ.

Heliotropium luteum, Poir. Wâdy Ghurundel (Elim), and other wâdies between that and Mount Sinai.

H. rotundifolium, Boiss. Wâdies Harûn and Zuweirah ; about Bîr es Sebâ.

H. arbainense, Fres. From 'Akabah to the Ghôr es Safieh in many places.

H. undulatum, Vahl. Wâdy Lebweh ; Wâdy el 'Attîyeh to Jebel Herteh ; 'Akabah.

Anchusa aggregata, Lehm. Gaza.

A. ægyptiaca, Linn. Bethlehem and Gaza.

A. Milleri, Willd. Wâdy Zuweirah to Bîr es Sebâ ; Jericho and Gaza.

Podonosma syriacum, Lab. Jebel Abu Kosheibeh and Mount Hor ; Jerusalem ; Bethlehem.

Onosma (? sp.), ? *O. giganteum*, Linn. Apparently this species, but withered, on Mount Hor.

Echium plantagineum, Linn. Wâdy Zuweirah to Gaza ; Jaffa.

Arnebia cornuta, Ledeb. Jericho.

A. linearifolia, D.C. Wâdy Zûweirah to Bîr es Sebâ.

Lithospermum tenuiflorum, Linn. Wâdy Zuweirah to Bîr es Sebâ ; Jericho.

L. callosum, Vahl. Gaza ; Wâdy Ghurundel (Elim).

Alkanna tinctoria, Linn. Ramleh to Jerusalem, in several places.

A. orientalis, Linn. Jebel Mûsa ; Mount Hor and Jebel Abu Kosheibeh ; Wâdy es Sheikh ; Jerusalem.

Asperugo procumbens, Linn. Jericho.

Trichodesma africanum, R. Br. Wâdies Lebweh and es Sheikh; 'Akabah to Wâdy Harûn; Ghôr es Safieh.

SOLANACEÆ.

Solanum nigrum, Linn. Gaza and Ghôr es Safieh. Var. *moschatum*, in Wâdy Ghurundel (Edom). Very variable.

S. coagulans, Forsk., *S. sanctum*, Linn. Ghôr el Feifeh and es Safieh.

Withania somnifera, Linn. Tell Abu Hareireh, near Gaza.

Lycium europæum, Linn. Debbet er Ramleh and Wâdy Harûn, abundant; 'Akabah, between Tell Abu Hareireh and Bîr es Sebâ.

L. arabicum, Schwf. Wâdies el 'Ain and Harûn.

Mandragora officinarum, Linn. Tell Abu Hareireh to Gaza; Jericho; near the Ghôr.

Datura stramonium, Linn. Apparently this species, but not perfect, at Gaza.

Hyoscyamus muticus, Linn. Wâdies Sudur and el 'Ain; Wâdy Ghuweir. The 'sekkaran' of the Arabs.

H. aureus, Linn. Wâdy el 'Ain; Mount Hor; Tell Abu Hareireh to Gaza; Jericho and Jerusalem.

SCROPHULARIACEÆ.

Verbascum sinaiticum, D.C. Jebel Mûsa; Wâdies Zelegah and el 'Ain; Mount Hor.

V. sinuatum, Linn. 'Akabah; Jebel Herteh and Wâdy Hessi; Wady 'Arabah and Ghôr es Safieh.

**Celsia parviflora*, Dcne. Jebel Mûsa; Ghôr es Safieh. Only known from Jebel Katharîna in Sinai. An addition to the flora of Palestine.

Anarrhinum pubescens, Fres. Deir el 'Arbain at Jebel Katharîna; Wâdies Lebweh and es Sheikh. Confined to Sinai so far as known.

Linaria floribunda, Boiss. Wâdy el 'Attîyeh to Jebel Herteh, south-west of 'Akabah. Not previously found in Sinai.

LINARIA FLORIBUNDA, Boiss.

Linaria elatine, Linn. Ghôr es Safieh.

**L. macilenta*, Dcne. Wâdy el 'Ain ; 'Akabah ; Wâdy Harûn. The last locality makes this rare species an addition to the flora of Palestine.

L. Hoelava, Forsk., sub *Antirrhino.* Jericho and Gaza.

L. micrantha, Cavan. Jericho. Noted in ' Northern Plains' in Palestine.

L. albifrons, S. et Sm. Jericho.

Antirrhinum orontium, Linn. Ghôr es Safieh ; Jericho.

**Scrophularia heterophylla*, Willd., *forma.* Wâdy Mûsa, Petra, and Mount Hor. Appears to be known only from Greece and its confines. An addition to the flora of Palestine.

S. xanthoglossa, Boiss. Sands at Gaza.

S. deserti, Del. Wâdy el 'Ain ; El Tâbâ, Wâdy 'Arabah.

S. variegata, M.B. ? Jericho.

S. canina, Linn. ? Gaza. And other indeterminable Scrophulariæ from Petra and elsewhere.

**Lindenbergia sinaica*, Dcne. Wâdy el 'Attîyeh to Jebel Herteh ; Wâdy Ghuweir ; Ghôr es Safieh. An addition to the flora of Palestine.

Veronica anagallis, Linn. Jerusalem, Jericho, and Bethlehem.

V. beccabunga, Linn. Jerusalem.

V. syriaca, R. et S. Jericho ; Deir el 'Arbain at Jebel Katharîna, in the monastery enclosure.

OROBANCHACEÆ.

Orobanche (*Phelipæa*) *ægyptica*, Pers. On Cucurbitaceæ, near 'Ain es Sultân, at Jericho. A very showy blue species, with the base of the corolla yellow. In flower in January.

O. cernua, Loefl. Ghôr es Safieh.

ACANTHACEÆ.

Blepharis edulis, Forsk., *Acanthodium spicatum*, Del. Wâdies Lebweh and es Sheikh ; Ghôr es Safieh.

GLOBULARIEÆ.

Globularia alypum, Linn., var. *G. arabica*, J. et Sp. Jebel Abu Kosheibeh.

VERBENACEÆ.

Lantana camera, Linn. Cultivated at 'Ayûn Mûsa.

Verbena officinalis, Linn. Ghôr es Safieh.

Vitex agnus-castus, Linn. Between Jericho and the Jordan.

LABIATÆ.

Lavandula stœchas, Linn. Medjel to Jaffa.

L. coronopifolia, Poir. Wâdy Nasb to 'Akabah, frequent; wâdies about Mount Hor.

Mentha sylvestris, Linn., β *stenostachya*, Boiss. Ghôr es Safieh; Jericho.

M. sylvestris, Linn., γ *lavandulacea*, Boiss. Jebel Mûsa; Deir el 'Arbain at Jebel Katharina.

Lycopus europæus, Linn. Ghôr es Safieh. Recorded from North Palestine.

Origanum maru, Linn., β *sinaicum*, Boiss. Jebel Mûsa; Mount Hor.

O. maru, ? *forma*. Bethlehem.

Thymus capitatus, Linn. Between Medjel and Jaffa; Solomon's Pools.

Satureia cuneifolia, Ten.? Bîr es Sebâ to Tell Abu Hareireh.

S. cuneifolia, forma. Wâdy Ghurundel (Edom).

Micromeria nervosa, Desf. Between Bâb el Wâd and Jerusalem.

**M. sinaica*, Bth. Jebel Katharina. Rocky ledges above the Tufileh river, near the Ghôr. An addition to the flora of Palestine.

M. juliana, Linn., β *myrtifolia*, Boiss. Above Bâb el Wâd, between Ramleh and Jerusalem.

M. serpyllifolia, M.B., β *barbata*, Boiss. Between Ramleh and Jerusalem.

Salvia triloba, Linn. Bâb el Wâd, between Ramleh and Jerusalem.

S. verbenacea, Linn. Frequent on the Judæan plain, from Wâdy Zuweirah to Gaza.

S. controversa, Ten. Bîr es Sebâ to Gaza.

S. viridis, Linn. Jericho.

S. ægyptiaca, Linn. Wâdy Harûn; Ghôr es Safieh, and thence to Gaza.

**S. deserti*, Dcne. 'Akabah ; Wâdy Harûn. An addition to the flora of Palestine.

Nepeta septemcrenata, Ehr. Jebel Mûsa ; Wâdy es Sheikh.

Marrubium alysson, Linn. ? Specimens imperfect. Bîr es Sebâ.

Stachys affinis, Fres. Wâdies Lebweh, Berrâh, and es Sheikh ; frequent.

Lamium amplexicaule, Linn. Gaza ; Jericho ; Bîr es Sebâ.

Ballota undulata, Fres. Wâdies Lebweh, es Sheikh, and Ghurundel (Edom). Bîr es Sebâ to Tell Abu Hareireh.

B schimperi, D.C. Wâdy el 'Ain.

*Phlomis aurea, Dcne. Jebel Mûsa ; Mount Hor. An addition to the flora of Palestine.

Eremostachys laciniata, Linn. Bîr es Sebâ to Tell Abu Hareireh.

Prasium majus, Linn. Gaza.

Ajuga iva, Schreb. Gaza ; Wâdy Zuweirah to Bîr es Sebâ ; Jericho.

A. chia, Poir., 8 *tridactylites*. Jebel Katharîna and Mount Hor.

Teucrium polium, Linn. Zibb el Baheir ; Wâdies Lebweh, es Sheikh, and el 'Ain ; Jebel Mûsa and Jebel Katharîna. A common labiate from about 2,500 feet upwards.

**T. sinaicum*, Boiss.. Wâdy el 'Ain ; Mount Hor and Petra. An addition to the flora of Palestine.

PLUMBAGINEÆ.

Statice thouini, Viv. Jericho.

S. pruinosa, Linn. 'Ayûn Buweirdeh in the 'Arabah.

PLANTAGINEÆ.

Plantago albicans, Linn. Wâdy Sudur in Sinai ; Gaza.

**P. Loeflingii*, Linn. Jebel Usdum ; Jericho. An addition to the flora of Palestine.

P. ovata, Forsk. Wâdy Hessi in Sinai ; Ghôr es Safieh.

P. lagopus, Linn. Bethlehem ; Jericho.

P. arabica, Boiss. Jebel Mûsa.

P. arenaria, W.K. Jericho.

CYNOCRAMBEÆ.

Cynocrambe prostrata, Gært., *Thelygonum cynocrambe*, Linn. Between Ramleh and Jerusalem.

SALSOLACEÆ.

Chenopodium album, Linn. Jericho.

C. murale, Linn. Gaza ; Jericho ; Ghôr es Safieh.

Atriplex hastatum, Linn. Ghôr es Safieh.

A. tataricum, Linn., ? *forma*. Ghôr es Safieh.

A. patulum, Linn. Ghôr es Safieh.

**A. alexandrina*, Boiss. Jebel Usdum ; Jericho. Hitherto found only in Tunis and Alexandria. An addition to the flora of Palestine.

A. crystallinum, Ehr. 'Akabah. A scarce Egyptian species, not found elsewhere previously.

**A. leucocladum*, Boiss. Wâdies Ghurundel (Elim), es Sheikh, el 'Ain, etc. ; 'Akabah ; Ghôr es Safieh ; el Tâbâ in the 'Arabah. An addition to the flora of Palestine.

A. halimus, Linn. Wâdy el 'Ain ; Ghôr es Safieh.

Salicornia fruticosa, Linn. Gaza.

S. herbacea, Linn. Ghôr es Safieh.

Suæda monoica, Forsk. Wâdy es Sheikh, 'Akabah ; El Tâbâ in the 'Arabah ; Ghôr es Safieh. Dries black.

S. asphaltica, Boiss. 'Ayûn Buweirdeh in the 'Arabah.

Suæda fruticosa, Linn., β *brevifolia*. Ghôr es Safieh ; Jericho.
Traganum nudatum, Del. Wâdy Ghurundel, on the east side of the 'Arabah.
*Salsola inermis, Forsk. Jebel Usdum ; Petra ; Gaza. This species has been found only near Alexandria, in Forskahl's old locality. An addition to the flora of Palestine.
S. tetragona, Del. Ghôr es Safieh and Jebel Usdum ; Wâdy Ghuweir in Edom.
**S. longifolia*, Forsk. Jebel Usdum. Known only from Egypt, where Forskahl originally discovered it. An addition to the flora of Palestine.
**S. fœtida*, Del. Jebel Usdum ; El Tâbâ in the 'Arabah. An addition to the flora of Palestine.
S. rigida, Pall. Mount Hor.
S. rigida, Pall., var. *tenuifolia*, Boiss. Jebel Usdum.
Noœa spinosissima, Moq. Tell Abu Hareireh to Gaza ; Jebel Usdum ; Mount Hor, summit.
Anabasis aphylla, Linn. ? Wâdy Lebweh. Named doubtfully.
A. articulata, Forsk. The commonest plant throughout Sinai to the Dead Sea. Less frequent from the Ghôr to Gaza.
**A. setifera*, Moq. 'Ayûn Mûsa ; Wâdies Ghurundel (Elim), es Sheikh and el 'Ain ; Jebel Usdum ; Tell Abu Hareireh to Gaza. An addition to the flora of Palestine.
[N.B.—The species of the above order have been almost entirely determined by M. Boissier.]

AMARANTACEÆ.

Albersia blitum, Kunth. Ghôr es Safieh : at a well in Wâdy Nasb. Not native (?).
A. caudatus, Linn. Ghôr es Safieh. Appeared amongst seeds brought from there, and grown by Mr. Burbidge.
Œrua javanica, Juss. Frequent in Sinai, from Wâdy Sudur to 'Akabah ; Ghôr es Safieh. Var. β *bovei*, Welb., was also gathered, as in Wâdy Lebweh.

Digera arvensis, Forsk. Ghôr es Safieh and Ghôr el Feifeh. D. alternifolia, Linn., an erect annual, is the only species in Boissier. The plant of the Ghôr is the decumbent perennial given in Forskahl's 'Flora Egyptiaco-Arabico,' at p. 65. An addition to the flora of Palestine.

POLYGONEÆ.

Calligonum comosum, Her. Above 'Akabah, westward from the Haj-road to Jebel Herteh. 'Ayun Buweirdeh to the Ghôr, and elsewhere in the 'Arabah.

Emex spinosus, Linn. Gaza and Jericho.

Rumex roseus, Linn. Jericho; Ghôr es Safieh and Petra.

R. vesicarius, Linn. ? Ghôr es Safieh. These two species are (if distinct) sometimes confounded.

R. obtusifolius, Linn. Gaza.

Polygonum equisetiforme, S. et S. Frequent on the Edomitic side of the 'Arabah, from Petra northwards; Ghôr es Safieh, and between Bîr es Sebâ and Gaza.

NYCTAGINEÆ.

Boerhaavia plumbaginea, Cav., *forma,* var. (?) *incisa.* 'Akabah.

*B. verticillata, Poir. Wâdy Harûn and Ghôr es Safieh; between Gaza and Tell Abu Hareireh. Fragments from Wâdies Ghurundel and Ghuweir appear to belong to this species, which is recorded only from Muscat in South-east Arabia, in the Oriental region, by Boissier. It occurs, however, in Nubia, Senegal and Abyssinia, according to Hooker's 'Flora Nigritiana,' which is a not unfrequent range for the desert plants. An addition to the flora of Palestine.

*B. repens, Linn. Ghôr es Safieh and Jericho. An Egyptian species found in tropical Arabia, and from Morocco to India. An addition to the flora of Palestine.

THYMELÆACEÆ.

*Daphne linearifolia, sp. nov. Fruticosa ramosa, ramulis tenuibus rubellis, glabris foliosis, foliis linearibus acuminatis utrisque uninerviis glabris sessilibus, floribus 10-15 brevissime pedicellatis capitatis terminali-

LORANTHUS ACACIÆ, Zu

bus sessilibus, pedicellis scabriJulo hispidis, perigoniis dense et breviter villosis pallido-flavis, lobis lanceolatis mucronatis, tubo 2-3 plo brevioribus, bacca rubra-fusca,

Frutex 4-6 pedalis. Folia 1½-2 pollicaris, 1 linea lata. Perigonium sub anthese 2 lineas longum.

Hab. Petra et ad basim Montis Aaronis prope Petram, circa 2,800-3,400 pedes supra Mare Mediterraneum.

Affinis D. acuminata, Boiss., 'Flor. Orient.,' vol. iv., pp. 104-108, et D. mucronata, Royle., Ind. III., p. 322, tab. 81, sed distinguenda præsertim a foliis linearibus fere gramineis, et a floribus minutis pallidis-flavis. [Plate XVI., fig. 2.]

Thymelæa hirsuta, Linn. Mount Hor and Petra; Bîr es Sebá.

T. ? sp. Tell Abu Hareireh.

LORANTHACEÆ.

Viscum cruciatum, Sieb. Near Jerusalem. Very like our mistletoe, except in the berries, which are red.

Loranthus acaciæ, Zucc. Abundant on acacias and nubk (Zizyphus), in a grove several miles in length on the east side of the 'Arabah, about fifteen miles north of 'Akabah. In Wâdy Ghuweir, on tamarisk and nubk; plentiful in Ghôr el Feifeh and Ghôr es Safieh and near Jericho, on nubk. This species may be dependent on the visits of the sunbird for the fertilization of its flowers. Specimens of this bird (Cinnyris Oseæ) were obtained with their long bills dusted with the pollen of the tubular flowers, which the bill is well fitted to probe. I may add that they usually occur together, as at Jericho, Ghôr es Safieh, and near El Tâbâ. The berries are red, cup-shaped and soft, with a single large hard seed.

EUPHORBIACEÆ.

*Euphorbia ægyptiaca, Boiss. Ghôr es Safieh. A tropical African species, ranging from Egypt, Nubia and Abyssinia, to the Cape de Verdes, and Senegal on the west, and Northern India on the east. An addition to the flora of Palestine.

E. cornuta, Pers. Frequent in Sinai, especially on the western side.

SOME ACCOUNT OF THE

Euphorbia helioscopia, Linn. Jericho ; Gaza to Jaffa.

E. exigua, Linn. Apparently this species; but, although flowering, the plants were not an inch in height, and very difficult to detect. They were probably seedlings. At Tell Abu Hareireh and Gaza.

E. aulacosperma, Boiss. Bethlehem.

E. peplus, Linn. Jericho.

E. peploides, Gou. Gaza.

E. chamæpeplus, Boiss., β *sinaica*, Boiss. Jericho.

E. terracina, Linn. Gaza.

E. paralias, Linn. Gaza.

Andrachne aspera, Spreng. Frequent in Sinai from Wâdy Lebweh to 'Akabah, and along the 'Arabah to the Ghôr ; Gaza.

Crozophora obliqua, Vahl. Wâdies Nasb, es Sheikh, and Ghôr es Safieh.

Mercurialis annua, Linn. Gaza ; Jaffa; Jericho and Ghôr es Safieh.

Ricinus communis, Linn. Ghôr es Safieh ; near Gaza, a little southwards.

URTICACEÆ.

Urtica dioica, Linn. Jericho.

U. urens, Linn., and *U. pilulifera*, Linn. Both were gathered, I believe, at Gaza; but they were not in flower.

Parietaria (? sp.), ? *P. judaica*, Linn. Jerusalem, etc.

Forskahlia tenacissima, Linn. Frequent in Sinai. Plentiful at 'Akabah, and up the 'Arabah to the Ghôr.

Ficus carica, Linn. Wâdy Ghuweir, a valley leading from north-eastern 'Arabah to the Shobek country in Edom.

F. sycomorus, Linn. Gaza ; Wâdy Ghurundel (Edom).

CUPULIFERÆ.

Quercus coccifera, Linn. Bâb el Wâd to Jerusalem, etc.

*Salix acmophylla, Boiss. (?) Wâdy Ghuweir ; Ghôr es Safieh, by the Tufileh river. In the absence of inflorescence, Mr. Oliver would

not name the species positively. There seems, however, to be no doubt about it. An addition to the flora of Palestine.

Populus euphratica, Oliv. 'Ayûn Buweirdeh in the 'Arabah; Ghôr es Safieh and el Feifeh; by the Jordan below Jericho. Leaves of various shapes, from linear or linear lanceolate to broadly ovate, rhomboid, sinuate, inciso-dentate, lobed, or quite entire. A single mature tree near the Dead Sea is of considerable size, with uniform foliage. A bush of this species (saf-saf), I believe, occurs near the spring on Jebel Mûsa; but I unfortunately omitted to bring home a specimen. According to the Arabs, who say it makes the best charcoal for gunpowder, it occurs in the valley between Jebel Mûsa and Jebel Katharîna, but is nearly all gone.

ARACEÆ.

Arum (? sp.), ? *A. dioscoridis*, Sibth. Gaza.

Arum (? sp.). Summit of Mount Hor.

Arisarum vulgare, Targ. Common about Gaza.

PALMÆ.

Hyphæne thebaica, Del. 'Akabah. Found also in Sinai at Tôr. 'Akabah is the northernmost limit of the range of this palm. No doubt, in former times, it ranged through Sinai more abundantly, and perhaps to the Dead Sea. It would be interesting to submit some sections of the sub-fossil palms found in the Ghôr to microscopical examination. Captain Burton found this palm very frequent in Midian. Its headquarters are in tropical Africa.

Phœnix dactylifera, Linn. Frequent in Sinai and up the 'Arabah, nearly to the Dead Sea, but not occurring in the Ghôr es Safieh; Wâdy Ghuweir. A single well-grown tree at Er Rîha or modern Jericho, in the plain of the Jordan, seems to have escaped the observation of travellers. A solitary tree in the convent at Mar Saba is said to bear a stoneless date. The date-palms of Sinai are usually very poor specimens.

TYPHACEÆ.

* *Typha angustata*, B. et C. Wâdy Ghurundel (Elim); Wâdy el 'Ain in two places. I feel sure I saw it in the Ghôr, but omitted to preserve specimens. An addition to the flora of Palestine.

IRIDACEÆ.

Iris (Xiphion) palæstinum, Baker. Tell Abu Hareireh to Gaza; Gaza to Jaffa, etc., common. Coming into flower the last week in December. Flowers sweet-smelling.

AMARYLLIDACEÆ.

Sternbergia macrantha, Gay. Mount Hor, especially about the summit. In flower early in December. A large showy yellow species.

Narcissus Tazetta, Linn. Between Medjel and Jaffa. Coming into flower in the first week of January.

Pancratium Sickenbergeri, A. et S. Debbet er Ramleh; Wâdies Lebweh and es Sheikh; in several places in the 'Arabah, from 'Akabah to 'Ayûn Buweirdeh. Flowering November and December, but without leaves. A beautiful white-flowered species found elsewhere only in the neighbouring Egyptian deserts. Numerous bulbs which I brought home of this and other species failed to flower with Mr. Burbidge. The requisite dry heat is difficult to supply.

COLCHICACEÆ.

Colchicum montanum, Linn. Abundant on the plain of Judæa, from the Ghôr to Gaza; near the summit of Mount Hor. In flower the second week of December. Flowers white or mauve.

C. Steveni, Kunth. I gathered flowers, which Mr. Oliver refers to this species, on the summit of Jebel Katharîna, 8,500–8,600 feet, on November 19th; summit of Mount Hor, in flower December 10th. Boissier gives maritime plains of Palestine, Cilicia, and Mersina for its known geographical range. Flowers white or faintly lilac, and appearing without leaves on Jebel Katharîna, one, two or three to the scape. On Mount Hor the flowers were decidedly smaller, and sprang from a leafy shoot. The latter is normally the case in this species; and possibly the Jebel Katharîna plant is distinct, and belongs to the section of the genus in which the flowers are autumnal, the leaves vernal.

Androcymbium (Erythrostictus) palæstinum, Boiss. Wâdy 'Arabah, and wâdies leading into it near the Ghôr; Gaza; from near Jerusalem to Jericho; Ghôr es Safieh. In flower December 13th, and afterwards.

PALESTINE EXPLORATION FUND PL. 8

W.H.Fitch del
C.H.Fitch lith

XIPHION PALÆSTINUM, BAKER

Hanhart imp

PANCRATIUM SICKEMBERGERI, *Asch & Schweinf*

LILIACEÆ.

Gagea reticulata, Pall. Gaza; Bethlehem; Jericho; Jerusalem. In flower January 4th.

Uropetalum erythræum, Webb. (?) Sinai and in the 'Arabah, frequent; withered, but I think this species.

Ornithogalum umbellatum, Linn. Tell Abu Hareireh to Gaza, and at Gaza. Gaza to Jerusalem frequently.

Urginea scilla, Sternh. Abundant in Judæa. Apparently this species throughout Mount Hor and the Edomitic escarpment, at a moderate elevation, from the Ghôr es Safieh to Wâdy Ghurundel, but only in leaf.

**U. undulata*, Desf. This plant flowered since my return under Mr. Burbidge's care, and Mr. Baker refers it to the present species. These bulbs were greedily sought after by wild boars. I found them sparingly in the Wâdy 'Arabah near the Ghôr, and again at Bîr es Sebâ, generally only broken remains. An addition to the flora of Palestine.

Muscari racemosum, Linn. Jerusalem to Jericho; in flower in the second week of January.

Bellevallia flexuosa, Boiss. Wâdies Ghuweir, Harûn, Mûsa, and Abu Kosheibeh; Tell Abu Hareireh; Ghôr es Safieh; Bethlehem, etc. A white variety occurs near Bethlehem. Slightly sweet-scented. In flower first week of December.

Asphodelus fistulosus, Linn. Wâdies Lebweh and Ghuweir; Ghôr es Safieh; Gaza and Jericho. In flower in the middle of December.

A. ramosus, Linn. Abundant in the Judæan plain, from Bîr es Sebâ to Gaza; about the Ghôr es Safieh, from a few hundred feet above the level of the Dead Sea upwards. In some places nearly as abundant as the squill. In flower the first week of January.

Allium sinaiticum, Boiss. Wâdy Harûn and Mount Hor, 'a very rare species, with the alternate filaments bearing three distinct cusps, as in A. ponum and A. sativum.' Flowered in the Trinity College Botanic Gardens, Dublin, since my return. Determined by Mr. Baker.

[N.B.—Bulbous plants of many other species in the four foregoing orders were brought home, and are almost all growing. Only in a few instances have they, however, thrown out flowers.]

ASPARAGACEÆ.

Asparagus aphyllus, Linn. Jebel Abu Kosheibeh; Gaza; Ramleh to Jerusalem. Recorded from Nazareth only.

A. acutifolius, Linn. Mount Hor; Mejdel to Jaffa. Recorded only from Lower Lebanon.

JUNCACEÆ.

Juncus subulatus, Forsk., *J. multiflorus*, Desf. Ghôr es Safieh.

J. acutus, Linn. Gaza; a stunted form.

J. maritimus, Lam., β. *arabicus*, A.B. Wâdies Sudur, Ghurundel (Elim), and el 'Ain; 'Akabah; in the 'Arabah and Ghôr es Safieh. The common form was seen also at Gaza and Ghôr es Safieh. The desert form acquires the whitened appearance characteristic of its flora, and is very large.

CYPERACEÆ.

*Cyperus lævigatus, Linn. Wâdy el 'Ain; Gaza. An addition to the flora of Palestine.

C. distachyos, All.; *C. lævigatus*, var. *junciformis*, Cav. Tell Abu Hareireh; Gaza.

*C. eleusinoides, Kunth. Frequent on the margins of the Arundo jungles, near the Dead Sea, in the Ghôr es Safieh. Recorded only from Afghanistan in the Oriental Region. A native of India, tropical Africa, and Australia. An addition to the flora of Palestine.

C. papyrus, Linn. Mr. Oliver could not be positive about this species, the specimens, owing to the season, being imperfect. The plant had not attained the full size of the papyrus, but in the floral characteristics they agree. The papyrus occurs at Lake Huleh, near Jaffa, and at Gennesaret; but not elsewhere indigenous nearer than Nubia. At Sicily it is believed to be anciently introduced. It occurred in the Ghôr sparingly with the last species.

Cyperus longus, Linn. Ghôr es Safieh; Tell Abu Hareireh.
C. rotundus, Linn. Ghôr es Safieh; Gaza.
C. schænoides, Griseb. Gaza.
Scirpus holoschænus, Linn. Jebel Mûsa; Wâdy el 'Attiyeh to Jebel Herteh. Not previously found in Sinai.
S. maritimus, Linn. Ghôr es Safieh; very variable.
Fimbristylis ferruginea, Linn. Ghôr es Safieh.
F. dichotoma, Rottb. Ghôr es Safieh.
Carex divisa, Huds. Ghôr es Safieh.
C. stenophylla, Wahl.; β. *planifolia*, Boiss. Summit of Mount Hor; from Wâdy Zuweirah to Bîr es Sebá and Tell Abu Hareireh. Not found hitherto south of Gaza.

GRAMINEÆ.

Panicum teneriffæ, Linn. Wâdies el 'Air, el 'Attiyeh, and 'Arabah; 'Akabah.
P. colonum, Linn. Ghôr es Safieh. Found hitherto at Sidon only in Palestine.
P. repens, Linn. Gaza.
P. turgidum, Forsk. Wâdy el 'Ain, and in the 'Arabah; Wâdy Zelegah; 'Akabah.
**P. molle*, *P. barbinode*, Trin., *forma.* Jericho. Not included in Boissier's 'Flora Orientalis.' An addition to the flora of Palestine.
**Pennisetum dichotomum*, Forsk. Wâdies Nasb and el 'Ain; Debbet er Ramleh; frequent in the 'Arabah. May be added to the flora of Palestine.
P. cenchroides, Pers. Wâdy Harûn, and others leading into the 'Arabah from Edom; Wâdy 'Arabah and Ghôr es Safieh; Tell Abu Hareireh to Gaza.
Imperata cylindrica, P. de B. Wâdies Zelegah and el 'Ain; Ghôr es Safieh. Mentioned by Canon Tristram from Gennesaret under synonym I. aurundinacea, Cyr.

Saccharum ægyptiacum, Willd. Wâdy Ghuweir. Known from Sidon and Beirût in Palestine.

Erianthus ravennæ, P. de B. Ghôr es Safieh.

Sorghum halepense, Pers. Ghôr es Safieh. [*S. vulgare*, P., cultivated.]

Andropogon foveolatus, Del. 'Akabah. Found in Egypt and S.E. Arabia, only in the Oriental region.

A. annulatus, Forsk. Ghôr es Safieh.

A. hirtus, Linn. Summit of Mount Hor; Tell Abu Hareireh; Gaza.

Elionurus hirsutus, Vahl.; *Cælorachis hirsutus*, Brongn. Common in Sinai, and extending to the edge of the Ghôr.

Phalaris minor, Retz. Jericho.

Aristida cærulescens, Derf. Wâdy el 'Attlyeh.

A. ciliata, Desf. Wâdies Ghurundel and 'Arabah; Ghôr es Safieh.

A. plumosa, Linn. Debbet er Ramleh; Wâdy el 'Attiyeh; 'Akabah, and along the 'Arabah to the Ghôr. A most beautiful little grass, with long feathery styles.

A. plumosa, Linn., β. *Haussknechtii?* In the 'Arabah near the Ghôr.

A. obtusa, Del. Wâdies Nasb, Sudur, and Ghurundel (el 'Ain); Debbet er Ramleh.

Piptatherum multiflorum, Beauv. Wâdy es Sheikh; wâdies on both sides of the 'Arabah and in the main valley.

*Sporobolus spicatus, Vahl. 'Ayûn Mûsa; near the mouth of a valley a few miles south of Gaza, apparently this species. An addition to the flora of Palestine. Not previously recorded north of Cairo, nor in Sinai or Palestine.

*Agrostis verticillata, Vill. Jericho, and near Bethlehem. An addition to the flora of Palestine.

Polypogon monspeliense, Desf. Wâdy el 'Ain; Ghôr es Safieh.

Avena sterilis, Linn. Jebel Abu Kosheibeh; Bethlehem.

*Danthonia Forskahlii, Vahl. Wâdy Nasb; Debbet er Ramleh; Wâdy 'Arabah, near the Ghôr.

Cynodon dactylon, Pers. Ghôr es Safieh ; Tell Abu Hareireh ; Ramleh to Jerusalem.

C. dactylon, var. Wâdy Ghurundel (Elim).

Phragmites communis, Trin. ; *P. gigantea*, J. Gay. Wâdies Ghurundel (Elim) and el 'Ain ; 'Ayûn Buweirdeh ; Wâdy Zelegah ; Ghôr es Safieh. This species appears to have been called Arundo donax by several writers, especially in Sinai. In Wâdy el 'Ain this gigantic species reached a height of 15 feet.

Lamarckia aurea, Moench. Jericho.

Koeleria phlæoides, Vill. Jericho.

**Eragrostis pæoides*, P. de B. Ghôr es Safieh. An addition to the flora of Palestine.

**E. pilosa*, Linn. Ghôr es Safieh. An addition to the flora of Palestine.

**E. megastachya*, Link. Ghôr es Safieh. An addition to the flora of Palestine.

E. cynosuroides, Retz. Abundant at 'Ain Tâbâ, near El Tâbâ, in the 'Arabah ; Ghôr es Safieh ; Gaza.

Schismus marginatus, P. de B. Jericho.

Poa annua, Linn. Ghôr es Safieh ; Tell Abu Hareireh.

P. ? sp., *P. sinaica*, R. ? Zibb el Baheir and Jebel Watlyeh, in Sinai.

Bromus madritensis, Linn. Jericho.

CONIFERÆ.

Pinus halepensis, Mill. Jerusalem.

Juniperus phœnicea, Linn. Jebel Abu Kosheibeh ; summit of Mount Hor ; plentiful along the brow of the Edomitic limestone escarpment east of the 'Arabah, where it appears to be the main ingredient of a respectable growth of trees. Doubtfully mentioned by Tristram from Lebanon, but not elsewhere in Palestine. The present locality is in Boissier, partly.

Juniperus, (?) sp. Bethlehem. Not in an identifiable condition.

Ephedra fragilis, Desf. Wâdy 'Arabah, between 'Ayûn Buweirdeh and the Ghôr; Jericho; near Bâb el Wâd; Bethlehem, where specimens were gathered with inflorescence.

E. elata, Dcne. 'Ayûn Buweirdeh, near the Ghôr. Apparently this species.

E. alte, C. A. Mey. Summit of Jebel Mûsa; Wâdy Hessi and Jebel Herteh.

EQUISETACEÆ.

**Equisetum elongatum*, Willd. Ghôr es Safieh; Gaza. Five or six feet high in the Ghôr.

FILICES.

Ceterach officinarum, Willd. Bâb el Wâd and Solomon's Pools, in Judæa; Jebel Abu Kosheibeh and Mount Hor, in Edom.

Notholæna lanuginosa, Desf. Wâdy Harûn and Jebel Abu Kosheibeh.

Cheilanthes fragrans, Linn. Jebel Abu Kosheibeh and Mount Hor; Bethlehem and Bâb el Wâd.

Adiantum capillus-veneris, Linn. Jebel Mûsa; Bethlehem.

CHARACEÆ.

*Chara hispida, Linn. 'Ayûn Buweirdeh in the 'Arabah, and at Bîr es Saura in Wâdy el 'Attîyeh.

MUSCI.[1]

Anisothecium (*Dicranum*) *varium*, Hedw. Gaza, barren.

Grimmia apocarpa, Linn. Jebel Mûsa, Jebel Katharîna, barren.

G. leucophæa, Grev. With the preceding, barren.

**G. trichophylla*, Grev. Jericho; a few slender stems with very young fruit.

**G. pulvinata*, Linn. Jericho.

**G. crinita*. Mount Hor; Wâdy Mûsa.

[1] These mosses have been determined by Mr. William Mitten, F.L.S.

Hymenostylium (*Gymnostomum*) *rupestre*, Schw. Jebel Mûsa, Jebel Katharina, barren.

H. (*Eucladium*) *verticillatum*, B. et S., *Weissia verticillata*, Brid. With the last, and barren.

*Tortula vinealis, Brid. Jericho, barren.

**T. unguiculata*, H. et T. Gaza. A few barren stems.

**T. revoluta*, H. et T. Gaza. Barren stems intermixed.

T. ruralis, Linn. Jericho, with fruit ; Bîr es Sebâ ; Jaffa. Piliferous.

T. membranifolia, Hook. Jericho ; Edom, with old fruit.

**T. inermis*, Mont. Jebel Mûsa and Wâdy Harûn.

T. ambigua, B. et S. Gaza ; Mount Hor.

**T.* (*Trichostomum*) *rigidula*, Hedw. Gaza, barren. Seems to be this species.

T. (*Trichostomum*) *tophacea*, Brid. Mount Hor, barren.

**T.* (*Trichostomum*) *nitida*, Lindb. Same locality as last.

**Encalypta vulgaris*, Hedw. Jebel Katharina and Mount Hor.

Entosthodon Templetoni, Schw. Jebel Mûsa and Jebel Katharina.

Funaria hygrometrica, Linn. Hospital of St. John, ruins ; Jerusalem.

*Bryum argenteum, Linn. Jericho.

**B. atropurpureum*, W. et Mohr. Jaffa ; Bîr es Sebâ ; Jericho. Chiefly barren.

B. turbinatum, Hedw. Jebel Mûsa ; Jebel Katharina. Some others which will not revive are barren and uncertain.

Hypnum (*Brachythecium*) *Ehrenbergii*, Lorentz. Jericho.

H. velutinum, L. Jebel Mûsa and Jebel Katharina.

H. ruscifolium, Neck. Jebel Katharina.

HEPATICÆ.

**Fossombronia angulosa*, Raddi. Gaza.

*Otiona aitonia, Corda. Jericho. Certainly of this genus, probably this species.

SOME ACCOUNT OF THE

Lunularia vulgaris. Hospital of St. John, Jerusalem.
Riccia lamellosa, Raddi. Bîr es Sebà.

[N.B.—The majority of the above mosses and hepaticæ have not been previously known to inhabit Sinai or Palestine. They are almost invariably common Mediterranean species, the greater part being found in Britain. Lorentz, in the 'Proceedings of the Berlin Academy,' 1867, enumerated twelve species of those collected by Ehrenberg, 1820-26, in Egypt and Sinai; and Mons. Barbey, in his 'Herborization au Levant,' Lausanne, 1882, gives a list of fifteen Palestine mosses and two hepatics, chiefly from Beyrout. Seven of Lorentz's and four of Barbey's are in my list. Decaisne, in the 'Flora Sinaica,' also gives four mosses from Sinai, of which two are not in my collection. Mr. Mitten was not aware of any Palestine mosses except those of Lorentz, and I have been able only to give the above further references. Altogether the total is forty-three mosses and six hepaticæ, but three of the mosses which Lorentz described as new may, according to Mr. Mitten, be found to be not so if good specimens were procured. Lorentz's specimens were all barren.

LICHENES.[1]

Omphalaria, sp. ? Sterile and uncertain. Jebel Katharîna, on rocks.
Ramalina crispatula, Nyl. Sterile. On the ground at Bîr es Sebà. Distr. Canaries and N. Africa.
Cladonia pyxidata, L. Sterile, and but little evolute. On the ground at Jericho. Cosmopolitan.
Lecanora (Squamaria) crassa, D.C. Sparingly fertile. Amongst mosses on rocks in Wâdy Ghurundel (Edom). Distr. Europe, Africa, and Australasia.
Lecanora (Sarcogyne) pruinosa, Sm. On rocks. Tîh escarpment, near Wâdy Zelegah. Distr. Europe, N America.
Endocarpon hepaticum, Ach. On the ground. Ghôr es Safieh. Distr. Europe, Africa, N. America.

[1] Several species from Wâdy Harûn do not appear in the above list, for which I am indebted to the Rev. James Crombie, F.L.S.

AN ANALYSIS OF THE FLORA OF SINAI

AND

GENERAL REMARKS ON ITS BOTANY, AND THAT OF THE DEAD SEA BASIN.

AN ANALYSIS OF THE FLORA OF SINAI.

DESCRIPTION OF SINAI.

THE Peninsula of Sinai in the Red Sea extends from N. lat. 30° to N. lat. 27° 40′ and from E. long. 32° 39′ to E. long. 35°. 'The Red Sea at its northern end divides into arms of unequal length. The eastern and shorter of these is the Gulf of 'Akabah, the western and longer one the Gulf of Suez, and the Peninsula of Sinai is the triangular promontory which lies between them. The base of this triangle is a line about 150 miles long drawn from Suez at the head of the one gulf to 'Akabah at the head of the other; its two sides measured from these points respectively to its apex at Râs Muhammed, the southernmost point of the promontory, are about 186 and 133 miles in length, and the area of the peninsula thus defined is about 11,500 square miles, or, roughly speaking, twice that of Yorkshire.'[1]

The north-eastern corner of this peninsula, 'Akabah, lies about half a degree south of the north-western at Suez.

The northern boundary of Sinai, as thus defined, corresponds almost exactly with the Haj route from Cairo, and cuts through the middle of the plateau known as the Desert of the Tîh. This elevated table-land, which is very distinct from the rest of Sinai, extends with a gradually increased elevation from the base of Sinai in a semicircular sweep to about the middle of the peninsula, and breaks down in a series of lofty and often inaccessible cliffs known as the Tîh escarpment. The part of

[1] 'Ordnance Survey of Sinai,' 1869, Part I., p. 18.

Missing Page

Missing Page

Nasturtium officinale, Rosa rubiginosa, Lamium amplexicaule, Scirpus Holoschœnus, Schœnus nigricans, which would most easily fall under the 'Mid-Europe' division; others, still fewer, as Echinospermum spinocarpos, Paracaryum micranthum, Plantago ovata, belong to the 'Aralo-Caspian,' and a more considerable number, of which may be instanced Matthiola oxyceras, Alsine picta, Holosteum liniflorum, Astragalus bombycinis, Zozimia absynthifolia, Achillea santolina, A. fragrantissima, Kochia latifolia, belong to the Mesopotamian group. Most Mesopotamian species, however, which range as far as Sinai have equal claims to be regarded as Mediterranean, or, in some cases, as Desert; and similarly a large number of plants, several of which appear in Sinai, may be distributed with equal justice to a position in any of the four subdivisions of the 'Oriental' region, their real requirement being that of saline soil.

In the following tables, a few species recorded from the 'Tih plateau,' or 'Arabia Petræa conterminous to Palestine,' are included with a mark of doubt, the locality being somewhere on the northern edge of the Peninsula, but possibly lying outside the boundary.

An asterisk is appended to those names which do not appear in the flora of Palestine list as given by Canon Tristram. A few probably introduced plants are bracketed.

Those printed in italic characters are peculiar to Sinai, so far as at present known. Eight species considered endemic have been found beyond these limits, six by me and two by Burton, since their first records in Boissier. The letters End. indicate the group to which the peculiar species belong.

The groups to which I have referred the flora are Desert, Mediterranean and Plateaux. The first two columns being devoted to the Desert species, which predominate, although if the North African flora be all regarded as Mediterranean, several of my Desert plants might be referred to that section.

The first column is devoted to those Desert species whose range is limited to the immediately adjoining countries of Palestine, Egypt and Arabia Petræa (of which Sinai is a part), and in some cases the Syrian desert. In many cases the plants of this column are almost as essentially Sinaitic as the peculiar species.

The second column includes Desert species with a wider range. Too

REMARKS ON TABLE OF FLORA OF SINAI.

large a space would be occupied by contractions representing accuracy in enumerating the lists of stations given by Boissier. I have therefore only indicated the *tendency* of the distribution of the species. Thus E. signifies that the species reaches Persia, Beloochistan, Scinde, Afghanistan, or even India on the East ; W. stands for North Africa, Canaries or Cape Verdes on the West ; and T. intimates that the plant is found either in or on the edge of the Tropics, chiefly in Nubia, Abyssinia and Arabia, Broadly speaking, E. generally stands for Persia and Beluchistan, and W. for North-west Africa and islands.

The third column includes Mediterranean species. I have here again endeavoured to tell something of the range of the plant by a system of letters. Thus an asterisk stands for typical Mediterranean ; L. for local Mediterranean, confined to a very limited portion of its coast ; E. shows that the species has a further wide range eastwards to Persia ; and W. that the plant inhabits the western countries of the Mediterranean, sometimes even reaching the Canaries, and thus illustrating the transition between the Mediterranean and the Desert floras. The transition on the tropical side is, however, more obvious.

The fourth column is the most difficult and uncertain one to deal with. The elevation at which the plant appears on Sinai itself has been taken as a first guide, but many Sinaitic species have a wide, vertical distribution, and those which have, will be found more often to be of a Mesopotamian or Syrian desert character, than either Mediterranean, Montane or Plateaux. Most of the Plateaux or Mountain plants of Sinai, which are properly so classed, reappear in the elevated parts of Persia (E.), a few in Songaria, Turkestan and the Caucasus (N.E.), some more in the mountains of Asia Minor, Taurus and the Lebanon range (N.), a very small number in the North-western Mediterranean mountains (N.W.), and the Western Algeria or Morocco Plateaux (W.), and two or three in the southern mountains of Nubia and Abyssinia (S.).

The last column is reserved for species too widespread to fall in line elsewhere. Tp. signifies that they are chiefly of the temperate region of the Eastern Hemisphere ; Wm. is placed after plants of warm or tropical countries, widely spread in the Old World ; while W.W. gives a further ' world-wide ' range to the Western Hemisphere.

TABULAR VIEW OF THE FLORA OF THE SINAITIC PENINSULA.

	DESERT		MEDIT.	PLATEAUX.	WIDE-SPREAD.	
	Sinaitic and adjoining regions	Wide-spread.				
Nigella deserti, Boiss.		E.				1
*(Papaver Decaisnei, Hochst.)				E.		2
Rœmeria orientalis, Boiss.				E., N.E.		3
*Glaucium arabicum, Fres.	END.					4
Cocculus Leæba, DC.		T.				5
*Morettia philœana, Del.		T.				6
M. canescens, Boiss.	*					7
Matthiola arabica, Boiss.	*					8
M. oxyceras, DC.			E.			9
M. livida, Del.	*					10
Eremobium lineare, Del.		E.				11
Farsetia ægyptiaca, Turr.		E.,W.				12
*F. ovalis, Boiss.		E.				13
Nasturtium officinale, R. Br.					W.W.	14
Sisymbrium Schimperi, Boiss.				E.		15
(S. pannonicum, Jacq. var.)					TP.	16
S. erysimoides, Desf.		E.W.T.				17
(S. irio, Linn.)					TP. WM.	18
Malcolmia pulchella, DC.	*					19
M. africana, Linn.					WM.	20
*M. aculeolata, Boiss.				E.		21
Nasturtiopsis arabica, Boiss.	*					22
*Alyssum marginatum, Stend.				E.		23
A. homalocarpum, F. M.				END.		24
Clypeola microcarpa, Moris.			E.			25
Notoceras canariense, R. Br.		E.,W.				26
Anastatica hierochuntina, Linn.		E.W.T.				27
(Lepidium chalepense, Linn.)				N.		28
(L. Draba, Linn.)					WM.	29
(Erucaria aleppica, Gærtn.)		LOC.				30
Hussonia uncata, Boiss.		W.				31
Schimpera arabica, Hochst.		E.				32
Moricandia arvensis, Linn.			*			33

TABULAR VIEW OF THE FLORA OF THE SINAITIC PENINSULA.

	DESERT.		MEDIT.	PLATEAUX.	WIDE-SPREAD.	
	Sinaitic and adjoining regions.	Widespread.				
Moricandia sinaica, Boiss		E.				34
M. dumosa, Boiss. .						35
*M. clavata, Boiss.		W.				36
Diplotaxis Harra, Forsk.		E.				37
D. acris, Forsk.	*					38
Brassica Tournefortii, Gou.		E.				39
(Sinapis arvensis, Linn.).					TP.	40
Savignya ægyptiaca, DC. et var.		F.				41
Schouwia Schimperi, J. et Sp.	*					42
? Enarthrocarpus strangulatus, Boiss.						43
(Raphanus sativus, Linn.)					TP.	44
Zilla myagroides, Forsk.	*					45
Cleome arabica, Linn.		W.				46
? *C. brachycarpa, Vahl.		E.				47
C. trinervia, Fres.	*					48
C. droserifolia, Del.	*					49
*C. chrysantha, Dcne.		T.				50
? *Mœrua uniflora, Vahl.		E.,W.,T.				51
Capparis spinosa, Linn.			* E.			52
C. galeata, Fres.	*					53
Ochradenus baccatus, Del.		E.				54
Reseda arabica		E.,W.				55
R. propinqua, R. Br.		W.				56
*R. stenostachya, Boiss.		(?) L.				57
R. muricata, Presl.	*					58
R. pruinosa, Del.	*					59
Oligomeris subulata, Del.[1]		F.,W.				60
Caylusea canescens, Linn.		W.				61
*Helianthemum ventosum, Boiss.				END.		62
H. kahiricum, Del.		W.				63
H. Lippii, Linn.			I..,E.			64
H. vesicarium, Boiss.			L.			65
*Polygala spinescens, Dcne.				END.		66
*Dianthus sinaicus, Boiss.				END.		67

[1] California and New Mexico.

AN ANALYSIS OF THE FLORA OF SINAI.

	DESERT.		MEDIT.	PLATEAUX.	WIDE-SPREAD.	
	Sinaitic and adjoining regions.	Wide-spread.				
(Saponaria vaccaria, Linn.)					TP.	68
Gypsophila Rokejeka, Del.			L., E. (?)			69
*G. elegans, MB.				N.E.		70
G. hirsuta, Lab. var. δ				N.,N.W.		71
Silene conoidea, Linn.			* E.			72
S. villosa, Forsk.			W.			73
*S. arabica, Boiss.	END.					74
*S. linearis, Dcne.	END.					75
*S. leucophylla, Boiss.				END.		76
*S. Schimperiana, Boiss.				END.		77
Spergularia diandra, Guss.			* E.			78
*Buffonia multiceps, Dcne.				END.		79
Alsine Meyeri, Boiss.				N.,NE.,E.		80
A. picta, S. et Sm. var. γ			L.,E.			81
Arenaria graveolens, Sch.				NW.N.NE.		82
Holosteum liniflorum, Stev.				NE.,E.		83
Robbairea prostrata, Forsk.		W.,E.,T.				84
Polycarpon arabicum, Boiss.	*					85
P. succulentum, Del.	*					86
Polycarpæa fragilis, Del.		W.				87
Herniaria cinerea, DC.			* E.,W.			88
H. hemistemon, J. Gay.		W.				89
Paronychia sinaica, Fres.	? *					90
P. desertorum, Boiss.		E.				91
Gymnocarpum fruticosum, Pers.		E.,W.				92
Sclerocephalus arabicus, Boiss.		E.,W.				93
Pteranthus echinatus, Desf.		E.,W.				94
*Cometes abyssinica, R. Br.		T.				95
*Telephium sphærospermum, Boiss.	*					96
Glinus lotoides, Linn.					WM.	97
Portulaca oleracea, Linn.					W.W.	98
*Reaumuria hirtella, J. et Sp.	*					99
Tamarix gallica, Fl. Gr. var.					TP. WM.	100
T. articulata, Vahl.		E.,W.,T.				101
Frankenia pulverulenta, Linn.					TP. WM.	102
Frankenia hirsuta, Linn. γ					TP.	103
*Hypericum sinaicum, Hochst.				END.		104

TABULAR VIEW OF THE FLORA OF THE SINAITIC PENINSULA. 131

	DESERT.		MEDIT.	PLATEAUX.	WIDE-SPREAD.	
	Sinaitic and adjoining regions.	Widespread.				
Malva rotundifolia, Linn.					TP.	105
M. parviflora, Linn.			* E.			106
*Althæa striata, DC.				END.		107
A. rosea, Linn.			L.			108
Abutilon fruticosum, G. et P.		E.,T.				109
A. muticum, Del.		E.,T.				110
Hibiscus ovalifolius, Vahl.		T.				111
Erodium laciniatum, Cav.			* E.			112
E. hirtum, Forsk.		W.				113
E. cicutarium, Linn.					TP.	114
E. bryoniæfolium, Boiss.			? E.			115
Monsonia nivea, Dcne.		W.				116
Tribulus terrestris, Linn.[1]			* E.			117
*T. bimucronatus, Viv.		E.,T.				118
T. alatus, Del.		E.,T.				119
Fagonia glutinosa, Del.[2]	*					120
F. kahirina, Boiss.		W.				121
F. Bruguieri, DC.		E.,W.				122
*F. myriacantha, Boiss.	END.					123
F. mollis, Del.	*					124
F. arabica, Linn.	*					125
Zygophyllum dumosum, Boiss.	*					126
Z. simplex, Linn.[1]		*				127
Z. album, Linn.			L.			128
Z. coccineum, Linn.		E.,T.				129
Z. decumbens, Del.	*					130
Seetzenia orientalis, Dcne.[1]		E.				131
Peganum Harmala, Linn.					TP. WM.	132
Nitraria tridentata, Desf.		W.,T.				133
Ruta tuberculata, Forsk.		W., E.,T.				134
Pistacia terebinthus, Linn.			L.			135
Zizyphus spina-christi, Linn.		E.,W.,T.				136
Rhamnus palæstina, Boiss.				N.		137
Moringa aptera, Gært.		T.				138
Crotalaria ægyptiaca, Benth.		E.				139

[1] Reappearing at the Cape of Good Hope.
[2] Mons. Boissier's discrimination of this genus is here followed. It would, perhaps, be safer to include these forms in one very variable *species*, F. cretica, Linn., as Anderson does.

17—2

AN ANALYSIS OF THE FLORA OF SINAI.

	DESERT.		MEDIT.	PLATEAUX.	WIDE-SPREAD.
	Sinaitic and adjoining regions.	Wide-spread.			
Lotononis dichotoma, Del.		E.,W.			140
L. persica, J. et Sp.		E.,T.			141
Argyrolobium uniflorum, Dcne.		W.			142
Retama Retam., Forsk.		W.			143
Ononis sicula, Guss.			L.,E.		144
(Trigonella fœnum - græcum, Linn.)			* T.,E.		145
T. stellata, Forsk.		W.			146
T. arabica, Del.	*				147
(Medicago tribuloides, Desv.).			*		148
M. laciniata, All.		E.,W.			149
(M. ciliaris, Willd.)			* W.		150
(Melilotus messanensis, Linn.)			* E.		151
(M. parviflora, Desf.)			* E.		152
Lotus lanuginosus, Vent.	*				153
L. arabicus, Linn.		T.,W.			154
Hippocrepis cornigera, Boiss.		E.,W.			155
Psoralea bituminosa, Linn.			* W.		156
? *Indigofera arabica, J. et Sp.		T.			157
*Tephrosia apollinea, Del.		T.			158
*T. purpurea, Pers.		T.			159
Colutea aleppica, Lam.				S.	160
Astragalus tribuloides, Del.		(?) *			161
*A. pseudostella, Del.			L.		162
*A. Schimperi, Boiss.	END.				163
? *A. eremophilus, Boiss.		E.,T.			164
A. tenuirugis, Boiss.		W.			165
A. bombycinus, Boiss.			E.		166
A. peregrinus, Vahl.		W.			167
*A. Fresenii, Dcne.			END.		168
A. sparsus, Dcne.	*				169
*A. acinaciferus, Boiss.	END.				170
*A. Sieberi, DC.	*				171
*A. trigonus, DC.	*				172
A. echinus, DC.				N.	173
A. Forskahlei, Boiss.	*				174
Onobrychis Ptolemaica, Del.	*				175

TABULAR VIEW OF THE FLORA OF THE SINAITIC PENINSULA. 133

	DESERT. Sinaitic and adjoining regions	DESERT. Widespread.	MEDIT.	PLATEAUX.	WIDE-SPREAD.	
Alhagi Maurorum, DC.			L.,E.			176
Cassia obovata, Coll.		E.,W.,T.				177
*C. acutifolia, Del.		E.,T.				178
Acacia nilotica, Del.		T.				179
A. tortilis, Haym.		T.,W.				180
A. Seyal., Del.		T.,W.				181
*Cratægus sinaica, Boiss.				END.		182
*Rosa rubiginosa, Linn. (var.)					TP.	183
Poterium verrucosum, Ehr.	*					184
Neurada procumbens, Linn.		E.,W.				185
Cucumis prophetarum, Linn.		E.,T.				186
Citrullus colocynthis, Linn.[1]		W.,E.,T.				187
Mesembryanthemum Forskahlei, Hochst.	*					188
Aizoon canariense, Linn.[1]		W.,E.,T.				189
Cotyledon umbilicus, Linn. var. ? .					(?) TP.	190
*Bupleurum linearifolium, DC. (γ)				NE.,E.		191
Apium graveolens, Linn.[1]					TP.	192
(Ridolfia segetum, Moris.)	*					193
Deverra tortuosa, Desf.	*					194
*D. triradiata, Hochst.				END.		195
Pimpinella cretica, Poir. β			L.			196
Coriandrum sativum, Linn.			L.			197
Scandix pinnatifida, Vent.				N.,NE.,E.		198
Pycnocycla tomentosa, Dcne.	*					199
*Ferula sinaica, Boiss.				END.		200
Zozimia absynthifolia, Vent.				N.,NE.,E.		201
*Oldenlandia Schimperi		T.				202
Crucianella membranacea, Boiss.	*					203
*C. ciliata, Lam. et β				N.		204
*Galium sinaicum, Dcne.				END.		205
(G. tricorne, With.)					TP.	206
G. spurium, Linn. γ				N.,NW.,E.		207
*G. Decaisnei, Boiss.				N.,E.,S.		208

[1] Found at the Cape of Good Hope.

AN ANALYSIS OF THE FLORA OF SINAI.

	DESERT.		MEDIT.	PLATEAUX.	WIDE-SPREAD.	
	Sinaitic and adjoining regions.	Widespread.				
Callipeltis cucullaria, Linn.					W.,N.,E.	209
(Cephalaria syriaca, Sch.)			* E.			210
*Scabiosa eremophila, Boiss.	END.					211
Pterocephalus sanctus, Dcne.				*		212
Erigeron Bovei, DC.	*					213
E. trilobum, Dcne.	*					214
Asteriscus pygmæus, C. et D.		E.,W.				215
A. graveolens, Forsk.		W.				216
*A. Schimperi, Boiss.	END.					217
? Anvillæa Garcini, Burm.		E.				218
Inula viscosa, Linn.				*		219
*Pulicaria longifolia, Boiss.		W.				220
P. undulata, Linn.		W.,T.				221
Francoeuria crispa, Forsk.		T.,E.,W.				222
Iphiona juniperifolia, Cass.	*					223
*I. scabra, DC.		T.				224
Varthamia montana, Vahl.				*		225
*Phagnalon nitidum, Fres.				E.		226
Lasiopogon muscoides, Desf.[1]			L.,E.			227
Leyssera capillifolia, Willd.		W.,T.				228
? Gymnarrhena micrantha, Desf.		E.,W.				229
Ifloga spicata, Forsk.[1]		E.,W.				230
? Achillea santolina, L.			? E.,L.			231
Santolina fragrantissima, Forsk.	(?) *					232
*Anthemis deserti, Boiss.	END.					233
*A. kahirica, Vis.	*					234
(Chrysanthemum coronarium, Linn.)			* W.			235
Pyrethrum auriculatum, Boiss.	*					236
*P. santalinoides, DC.	END.					237
Brocchia cinerea, Del.		W.,T.				238
Artemisia monosperma, Del.		L.				239
A. herba-alba, Asso. et vars.		S.,E.				240
A. judaica, Linn.	*					241
Senecio Decaisnei, DC.[1]		E.,W.				242

[1] Reappearing at the Cape of Good Hope.

TABULAR VIEW OF THE FLORA OF THE SINAITIC PENINSULA.

	DESERT.		MEDIT.	PLATEAUX.	WIDE-SPREAD.	
	Sinaitic and adjoining regions.	Wide-spread.				
Senecio coronopifolius, Desf. .			L., E.			243
Calendula arvensis, Linn. .					TP. WM.	244
Tripteris Vaillantii, Dcne. .	*					245
Echinops glaberrimus, DC. .	*					246
Carduus pycnocephalus, Jacq. γ					TP. WM.	247
Atractylis flava, Desf. .			E., W.			248
Onopodon ambiguum, Fres. .	(?) *					249
Phæopappus scoparius, Sieb. .	*					250
Amberboa Lippii, Linn. .			W.			251
A. crupinoides, Desf. .			W., T.			252
Centaurea eryngoides, Lam. .				N.		253
C. sinaica, DC. . . .	*					254
C. ægyptiaca, Linn. .			E.			255
Leontodon hispidulum, Del. .			E.			256
L. arabicum, Boiss. . .	*					257
Picris cyanocarpa, Boiss. .	*					258
Urospermum picroides, Linn. .			* E.			259
Zollikoferia mucronata, Forsk.			W., E.			260
? Z. tenuiloba, Boiss. . .	*					261
Z. arabica, Boiss. . .			W.			262
Z. nudicaulis, Linn. .			W.			263
*Z. fallax, J. et Sp. .			E., T.			264
? Z. massavensis, Fres. .			T.			265
*Z. glomerata, Cass. .			T., E., W.			266
Z. spinosa, Forsk. . .			W., E.			267
Picridium tingitanum, Linn. .			T., E., W.			268
Sonchus oleraceus, Linn. .					W.W.	269
Lagoseris bifida, Vis. .			L.			270
Campanula dulcis, Dcne. .				N.		271
Anagallis arvensis, Linn., α et β					W.W.	272
*Primula Boveana, Dcne. .				END.		273
Salvadora persica, Gærtn. .			E., W., T.			274
*Periploca aphylla, Dcne. .			E.			275
Solenostoma Argel., Del. .			T.			276
Calotropis procera, Willd. .			E., W., T.			277
Dæmia cordata, R. Br. . .			E., W., T.			278
Gomphocarpus sinaicus, Boiss.	*					279

	DESERT.		MEDIT.	PLATEAUX.	WIDE-SPREAD.	
	Sinaitic and adjoining regions.	Widespread.				
Leptadenia pyrotechnica, Forsk.	E.,T.					280
*Boucerosia sinaica, Dcne.			END.			281
Erythræa ramosissima, Pers.					TP.	282
E. spicata, Pers.					TP. WM.	283
Convolvulus hystrix, Vahl.	E.,T.					284
C. lanatus, Vahl.	*					285
C. arvensis, Linn.					W.W.	286
Cressa cretica, Linn.					W.W.	287
Cuscuta arabica, Fres.	*					288
Heliotropium luteum, Poir.	W.					289
H. arbaniense, Fres.	E.,T.					290
H. undulatum, Vahl.	W.					291
*H. persicum, Lam.	E.,T.					292
Anchusa hispida, Forsk.	W.,T.,E.					293
A. Milleri, Willd.	*					294
*Echium longifolium, Del.	(? W.) T.					295
*E. Rauwolfii, Del.	T.					296
Echiochilon fruticosum, Desf.	W.,T.					297
Arnebia hispidissima, Spreng.	E.,T.					298
A. cornuta, Ledeb.			E.,NE.			299
A. linearifolia, DC.			N.,E.			300
Lithospermum callosum, Vahl.	W.,E.					301
Alkanna orientalis, Boiss.			N.,E.			302
Echinospermum spinocarpos, Forsk.			N.,E.			303
*E. sinaicum, DC.			E.			304
*Paracaryum micranthum, Boiss.			S.,NE.,E.			305
P. rugulosum, DC.			NE.,E.			306
(Asperugo procumbens, Linn.)					TP.	307
Trichodesma africanum, Linn.[1]	E.,W.					308
Solanum nigrum, Linn. var.					W.W.	309
Withania somnifera, Linn.[1]	W.,E. *					310
Lycium europæum, Linn.	* W.					311
L. arabicum, Schw.	T.,E.					312

[1] Appearing again at the Cape of Good Hope.

TABULAR VIEW OF THE FLORA OF THE SINAITIC PENINSULA. 137

	Sinaitic and adjoining regions.	DESERT. Widespread.	MEDIT.	PLATEAUX.	WIDE-SPREAD.	
Hyoscyamus pusillus, Linn.				E.,NE.		313
H. muticus, Linn.		E.				314
Verbascum sinaiticum, Bth.				N.		315
V. sinuatum, Linn.			*E.,W.			316
*V. Schimperianum, Boiss.				END.		317
Celsia parviflora, Dcne.	*					318
*Anarrhinum pubescens, Fres.				END.		319
Linaria floribunda, Boiss.	*					320
(L. spuria, Willd.)			W.,E.*			321
L. ægyptiaca, Linn.		W.				322
L. macilenta, Dcne.		T.				323
Scrophularia deserti, Del.		E.				324
S. variegata, M.B.				N.,E.		325
Lindenbergia sinaica, Dcne.	*					326
Veronica Anagallis, Linn.					W.W.	327
V. Beccabunga, Linn.					TP.	328
V. camplopoda, Boiss.				NE., N., E.		329
Phelipæa tubulosa, Schenk.		E.				330
Blepharis edulis, Forsk.		T.,E.				331
Globularia arabica, J. et Sp.				*		332
*Lavandula pubescens, Dcne.		E.,T.				333
L. coronopifolia, Poir.		E.,T.				334
Mentha sylvestris, Linn. (γ, β).					TP.	335
Origanum Maru, Linn., β			L.			336
*Thymus decussatus, Bth.				END.		337
*Micromeria sinaica, Bth.				S.		338
Salvia palæstina, Bth.			L.,E.			339
S. controversa, Ten.			L.			340
S. ægyptiaca, Linn.		E.,W.,T.				341
*S. deserti, Dcne.				*		342
*Nepeta septemcrenata, Ehr.	*					343
Stachys affinis, Fres.	*					344
Lamium amplexicaule, Linn.					TP.	345
Ballota undulata, Fres.			L.			346
*Otostegia Schimperi, Bth.				END.		347
*O. moluccoides, Vahl.				S.		348
Leucas inflata, Bth.		T.,E.				349
Phlomis aurea, Dcne.				*		350

	DESERT		MEDIT.	PLATEAUX	WIDE-SPREAD	
	Sinaitic and adjoining regions	Wide-spread				
Ajuga chia, Poir. δ				E.,N.		351
*Teucrium leucocladum, Boiss.				END.		352
T. sinaicum, Boiss.				*		353
Statice Thouini, Viv.			L.,E.			354
S. pruinosa, Linn.			L.			355
Plantago albicans, Linn.			* E.			356
*P. cylindrica, Forsk.	*					357
*P. amplexicaulis, Cav.		E.,W.				358
P. ovata, Forsk.		E.,W.				359
P. ciliata, Desf.		(?W)E.T				360
P. arabica, Boiss.	*					361
P. phæostoma, Boiss.	*					362
Beta maritima, Linn.					TP.	363
Chenopodium murale, Linn.					W.W.	364
C. album, Linn.					W.W.	365
Atriplex dimorphostegium, K. et K.				W.,E.,NE.		366
A. crystallinum, Ehr.	*					367
A. leucocladum, Boiss.	*					368
A. halimus, Linn.			*			369
A. farinosum, Forsk.		T.,E.				370
Chenolea arabica, Boiss.	*					371
Kochia muricata, Linn.		E.,W.,T.				372
K. latifolia, Fres.			L.,E.			373
Arthrocnemum glaucum, Del.	*					374
*Halocnemum strobilaceum, Pall.					TP.	375
Suæda monoica, Forsk.		E.,T.				376
S. vermiculata, Forsk.		E.,W.,T.				377
*Schanginia baccata, Forsk.		E.,T.				378
Traganum nudatum, Del.		W.				379
Haloxylon articulatum, Cav.			L.			380
Salsola tetragona, Del.		W.				381
S. fœtida, Del.		T.,E.,W.				382
Anabasis articulata, Forsk.		W.				383
A. setifera, Moq.		E.				384
Halogeton alopecuroides, Del.	*					385
*(Amarantus caudatus, Linn.)		T.,E.				386

TABULAR VIEW OF THE FLORA OF THE SINAITIC PENINSULA. 139

	DESERT. Sinaitic and adjoining regions.	DESERT. Wide-spread.	MEDIT.	PLATEAUX.	WIDE-SPREAD.	
(Amarantus sylvestris, Desf.)					TP.	387
(Albersia blitum, Kunth.)					TP.	388
Ærua javanica, Juss.		W.,E., T.				389
Calligonum comosum, L'Her.		W.,E.				390
Rumex dentatus, Linn.		E.				391
*R. vesicarius, Linn.		E.,W.				392
R. roseus, Linn.			L.,E.			393
Atraphaxis spinosa, Linn. var.				NE.,E.		394
Boerhavia plumbaginea, Cav.[1]					WM.	395
B. verticillata, Poir.		T.,E.,W.				396
? Thymelæa hirsuta, Linn.			*			397
Euphorbia granulata, Forsk.		T.,E.,W.				398
E. cornuta, Pers.		W.				399
(E. peplus, Linn.)			*		TP.	400
E. chamæpeplus, Boiss.			*			401
*E. obovata, Dcne.	END.					402
E. terracina, Linn.		* W.				403
(E. helioscopia, Linn.)					TP.	404
Andrachne aspera, Spr.		T.,E.				405
Crozophora tinctoria, Linn.		* E.				406
C. obliqua, Vahl.		T.,E.,W.				407
Parietaria alsinefolia, Del.		E.				408
Forskahlea tenacissima, Linn.		E.,W., T.				409
Ficus pseudosycomorus, Dcne.	*					410
Salix safsaf, Forsk.		T.				411
? *S. babylonica, Linn.			E.			412
Populus euphratica, Oliv.			(?) E.			413
*Haplophila ovalis, R. Br.[2]						414
*H. stipulacea, Forsk.[2]						415
Ruppia rostellata, Koch.[2]						416
*Cymodocea rotundata, Ehr.[2]						417
*C. isoetifolia, Asch.[2]						418
*C. ciliata, Forsk.[2]						419
*Halodule uninervis, Forsk.[2]						420
*Hyphæne Thebaica, Del.		T.				421

[1] Appears again at the Cape of Good Hope.
[2] Seashore plants of Indian, Pacific, and Red Seas. Ruppia alone is found in the Mediterranean also.

AN ANALYSIS OF THE FLORA OF SINAI.

	DESERT		MEDIT.	PLATEAUX.	WIDE-SPREAD.	
	Sinaitic and adjoining regions.	Widespread.				
Phœnix dactylifera, L'Hert.	T.,E.,W.					422
*Typha angustata, B. et Ch.					WM.	423
Pancratium Sickembergeri, A. et S.	*					424
Colchicum montanum, Linn.				W.,N.,NE.		425
C. Steveni, Kunth.			L.			426
C. Ritchii, R. Br.			L.,W.			427
Erythrostictus palæstinus, Boiss.			L.			428
Gagea reticulata, Pall.					TP.	429
Allium sinaiticum, Boiss.	*					430
A. stamineum, Boiss.				N.,E.		431
*Uropetalum erythræum, Webb.	*					432
Asphodelus tenuifolius, Cav.					WM.	433
*A. pendulinus, C. et D.		W.				434
Asparagus stipularis, Forsk.			*			435
Juncus effusus, Linn.					W.W.	436
J. maritimus, Lam. (β)					TP.	437
J. punctorius, Linn.[1]				N.,S.,E.		438
*J. paniculatus, Hop.			L., ? E.			439
J. bufonius, Linn.					W.W.	440
Cyperus lævigatus, Linn.					W.W.	441
C. distachyus, All.			* E.			442
C. conglomeratus, R. et H.C.[2]	T.,E., W.					443
Scirpus holoschœnus, Linn.					TP.	444
S. maritimus, Linn.					W.W.	445
Schœnus nigricans, Linn.					TP.	446
? Carex divisa, Huds.					TP.	447
C. distans, Linn.					TP.	448
Panicum teneriffæ, Linn.		E.,W.				449
P. turgidum, Forsk.		T.,E.				450
Setaria glauca, Linn.					W.W.	451
? S. verticillata, Linn.					W.W.	452

[1] Found at the Cape of Good Hope.
[2] Found in Madagascar.

TABULAR VIEW OF THE FLORA OF THE SINAITIC PENINSULA.

	DESERT.		MEDIT.	PLATEAUX.	WIDE-SPREAD.	
	Sinaitic and adjoining regions	Widespread.				
Pennisetum dichotomum, Forsk.		E.,W.				453
*P. elatum, Hochst. . .	END.					454
P. ciliare, Link.[1] . .		E.,W.,T.				455
P. orientale, Rich. . .				N.,E., NE.		456
Imperata cylindrica, Linn. .					WM.	457
Saccharum ægyptiacum, Will.		T.,E.				458
Andropogon foveolatus, Del.[2]		T.,E.,W.				459
A. hirtus, Linn.[1] . .		T.,E.,W.				460
A. distachyus, Linn. . .			* W.			461
Elionurus hirsutus, Vahl. .		T.,E.				462
Phalaris minor, Retz.[1] . .			E., * W.			463
Aristida cærulescens, Desf.[1,2]		T.,E.,W.				464
A. pumila, Dcne. . . .		W.				465
*A. obtusa, Del.[1] . . .		W.,T.				466
A. ciliata, Desf.[1] . .		W.,T.				467
A. plumosa, Linn. . . .		W.				468
*A. hirtigluma, Steud. .		T.				469
? *A. acutiflora, Trim. .		W.,T.				470
*A. pungens, Desf. . .				(?) NE.		471
Stipa barbata, Desf. . .				E.,NE.,N.		472
Piptatherum multiflorum, Beauv.			* W.			473
*Sporobolus spicatus, Vahl. .		E.,T.				474
Agrostis verticillata, Vill. .					W.W.	475
Polypogon maritimum, Willd.					TP.	476
P. monspeliense, Linn. .					W.W.	477
Trisetum lineare, Forsk. .			L.			478
T. pumilum, Desf. . .			L.,W., E.			479
(Avena sterilis, Linn.) . .			* E.			480
(A. barbata, Brot.). . .			W. *			481
*Danthonia Forskahlei, Vahl.		W.,E.				482
Cynodon dactylon, Linn. .					W.W.	483
*Tetrapogon villosus, Desf. .		T.,W.,E.				484
(Eleusine indica, Linn.) . .					W.W.	485

[1] Found at the Cape of Good Hope.
[2] Found in Australia.

AN ANALYSIS OF THE FLORA OF SINAI.

	DESERT		MEDIT.	PLATEAUX	WIDE-SPREAD	
	Sinaitic and adjoining regions.	Widespread.				
Pappophorum brachstachyum, J. et Sp.	T., W., E.					486
*Boissiera bromoides, Hochst.				NE., E., N.		487
Phragmites communis, Trim. (β)					W.W.	488
Koeleria phleoides, Vill.					TP.	489
*K. sinaica, Boiss.				(?) END.		490
Eragrostis poæoides, P. de B.					TP.	491
Æluropus littoralis, Willd.					TP. WM.	492
Schismus arabicus, Nees.			L., E.			493
*Poa sinaica, Steud.				E., NE.		494
*P. soongarica, Sch.				E., NE.		495
*Vulpia myuros, Gmel.					W.W.	496
Scleropoa memphitica, Spr.			* E.			497
Bromus tectorium, Linn.					TP.	498
B. madritensis, Linn.					TP.	499
Brachpodium distachyum, Linn.			* E., W.			500
Ægopyrum squarrosum, Roth.				NE., E., W.		501
(Lolium multiflorum, Gaud.)					TP.	502
Lepturus incurvatus, Linn.					TP.	503
Ephedra alte, C. A. M.			L.			504
*E. alata, Dcne.				E., W., NE.		505
Nothochlæna lanuginosa, Desf.[1]			* E., W.			506
Cheilanthes fragrans, Linn.			*			507
Adiantum capillus-veneris, Linn.					W.W.	508
Equisetum ramosum, Schl.					W.W.	509

[1] Reappears in Australia.

ANALYSIS OF ORDERS REPRESENTED IN THE SINAITIC PENINSULA.

D.=DESERT, P.=PLATEAUX OR MONTANE, MED.=MEDITERRANEAN.

ORDERS AND NO. OF GENERA.	SPECIES.	D.	MED.	P.	ENDEMIC.	WIDE-SPREAD.
Papaveraceæ, 4	4	2		2	1 D.	
Menispermaceæ, 1	1	1				
Cruciferæ, 25	40	21	6		1 P.	18
Capparideæ, 3	8	7	1			
Resedaceæ, 4	8	7	1			
Cistineæ, 1	4	1	2	1	1 P.	
Polygaleæ, 1	1			1	1 P.	
Caryophylleæ, 9	17	3	4	9		1
Paronychieæ, 9	12	11	1			
Mollugineæ, 2	2	1				1
Portulaceæ, 1	1					1
Tamariscineæ, 2	3	2				1
Frankeniaceæ, 1	2					2
Hypericineæ, 1	1			1	1 P.	
Malvaceæ, 4	7	3	2	1	1 P.	1
Geraniaceæ, 2	5	2	1	1		1
Zygophylleæ, 6	17	14	2		1 D.	1
Rutaceæ, 1	1	1				
Terebinthaceæ, 1	1		1			
Rhamneæ, 2	2	1		1		
Moringeæ, 1	1	1				
Leguminosæ, 19	43	30	9	4	2 D., 1 P.	
Rosaceæ, 4	4	1	1	1	1 P.	1
Cucurbitaceæ, 2	2	2				
Ficoideæ, 2	2	2				
Crassulaceæ, 1	1					1
Umbelliferæ, 10	11	2	3	5	2 P.	1
Rubiaceæ, 4	8	2		5	1 P.	1
Dipsaceæ, 3	3	1	1	1	1 D.	
Compositæ, 37	58	43	9	3	3 D.	3
Campanulaceæ, 1	1			1		
Primulaceæ, 2	2			1		1
Salvadoraceæ, 1	1	1				1
Asclepiadeæ, 7	7	6		1	1 P.	
Gentianeæ, 1	2					2

AN ANALYSIS OF THE FLORA OF SINAI.

ORDERS AND NO. OF GENERA.	SPECIES.	D.	MED.	P.	ENDEMIC.	WIDE-SPREAD.
Convolvulaceæ, 3	5	3				2
Boragineæ, 11	20	12		7		1
Solanaceæ, 4	6	3	2	1		1
Scrophularieæ, 7	15	6	2	5	2P.	2
Orobanchaceæ, 1	1	1				
Acanthaceæ, 1	1	1				
Globularieæ, 1	1			1		
Labiatæ, 15	21	6	4	9	3P.	1
Plumbagineæ, 1	2		2			
Plantagineæ, 1	7	6	1			
Salsolaceæ, 14	23	14	4	1		4
Amarantaceæ, 3	4	2	2			
Polygoneæ, 3	5	3	1	1		
Nyctagineæ, 1	2	1	1			
? Thymelæaceæ, 1	1		1			
Euphorbiaceæ, 3	10	6	2	1		1
Urticaceæ, 3	3	3				
Salicineæ, 2	?3		1		2	
Hydrocharideæ, 1	2	} Marine.				
Potameæ, 3	5					
Palmæ, 2	2	2				
Typhaceæ, 1	1	1				
Amaryllidaceæ, 1	1	1				
Colchicaceæ, 1	3		2	1		
Liliaceæ, 5	7	3	1	1		2
Asparagaceæ, 1	1		1			
Juncaceæ, 1	5		1	1		3
Cyperaceæ, 4	8	1		1		6
Gramineæ, 35	55	20	10	8	2D.	17
Gnetaceæ, 1	2		1	1		
Filices, 3	3		2			1
Equisetaceæ, 1	1			1		
Characeæ, 1	1			1		
Musci, 9	?15				1	?14

LIST OF ORDERS ARRANGED ACCORDING TO THE NUMBER OF THEIR SINAITIC SPECIES.

Tamariscineæ, Dipsaceæ, Urticaceæ, Salicineæ, Colchicaceæ, and Filices three each. Eleven orders have but two representatives, while nineteen orders have but a single species appearing in Sinai. The average is about seven and a half species per order.

Compositæ	... 58	Paronychieæ	... 12	Solanaceæ	... 6
Gramineæ	... 55	Umbelliferæ	... 11	Geraniaceæ	... ⎫
Leguminosæ	... 43	Euphorbiaceæ	10	Convolvulaceæ	⎬ 5
Cruciferæ	... 40	Resedaceæ	... ⎫	Polygoneæ	... ⎭
Salsolaceæ	... 23	Capparidaceæ	⎬ 8	Juncaceæ	... ⎫
Labiatæ	... 21	Rubiaceæ	... ⎭	Papaveraceæ	... ⎪
Boragineæ	... 20	Cyperaceæ	... ⎫	Cistineæ	... ⎬ 4
Caryophylleæ	17	Malvaceæ	... ⎪	Rosaceæ	... ⎪
Zygophylleæ	... 17	Asclepiadaceæ	⎬ 7	Amarantaceæ	⎭
(Musci)	... 15	Plantagineæ	... ⎪		
Scrophularieæ	15	Liliaceæ	... ⎭		

In the foregoing list the following genera have no representatives in Europe:

Cocculus.	Sclerocephalus.	Anvillæa.	Lindenbergia.
Morettia.	Cometes.	Francoeuria.	Blepharis.
Eremobium.	Abutilon.	Iphiona.	Otostegia.
Nasturtiopsis.	Seetzenia.	Varthamia.	Leucas.
Anastatica.	Moringa.	Gymnarrhena.	Chenolea.
Hussonia.	Crotalaria.	Brocchia.	Tragonum.
Schimpera.	Lotononis.	Tripteris.	Haloxylon.
Savignya.	Indigofera.	Salvadora.	Ærua.
Schouwia.	Tephrosia.	Solenostoma.	Haplophila.
Zilla.	Cassia.	Calotropis.	Halodule.
Mœrua.	Acacia.	Dæmia.	Phœnix.
Ochradenus.	Neurada.	Leptadenia.	Hyphæne.
Oligomeris.	Citrullus.	Boucerosia.	Elionurus.
Caylusea.	Deverra.	Arnebia.	Pappophorum.
Robbairea.	Pycnocycla.	Echiochilon.	Boissiera.
Polycarpæa.	Zozimia.	Paracaryum.	Ægopyrum.
Gymnocarpum.	Oldenlandia.	Trichodesma.	

These form a little less than a quarter of the total number of genera in the peninsula, deducting a few from that total of those included in brackets, which are undoubtedly not native.

If we omit from consideration the ubiquitous species (74) classed in the last column, there will be found to be a very small proportion of species common to Sinai and Europe, almost all of which belong, of course, to the group classed as essentially Mediterranean. In round numbers, the Sinaitic flora is almost exactly 500, of which less than a third is common to Europe. Of this 150 species or thereabouts, at least twenty have almost certainly been introduced by human agency from Europe to Sinai, and a large proportion of the remainder are so widespread as to afford now but slight clue to their original 'centre of dispersion.' We must, therefore, seek some other flora to help us to study that of Sinai.

If the European flora was found to form any considerable portion of that of Sinai, it would be probably in the Plateaux region, which is, however, conspicuously non-European, as will be presently seen.

DESERT FLORA.

If we consider the first column of Desert species, we shall find eight peculiar to Sinai, of which several belong to a slightly more elevated region than that which they must inhabit before they can extend their range in existing circumstances across the low-lying desert strips which isolate Sinai from neighbouring countries. They hardly, however, pertain to the Plateaux region. Of the remainder, twenty-six do not reach Egypt, but are confined to Sinai and the immediately adjacent parts of Palestine and Arabia Petræa, as Edom, Judæa, Midian or Mount Seir. These may have originated in Sinai, and as yet, since its union to Africa, spread no further west. I incline to think these are often of modern development, due to a climate which has only arrived at its present peculiarities since the glacial epoch. The rest of this column, about forty-five in number, are species extending westwards to Egypt, but otherwise localized either in Sinai only or as above, with the exception of a very few occurring also in the Syrian Desert. Viewed in the above light of recent development, and more especially in connection with the many forms to which some genera, as Fagonia, Tamarix, Reseda, Reaumuria, Astragalus, Zollikoferia, Salsolaceæ, Andropogon, and others, are (to the confusion of botanists) giving rise, this highly modified flora may yet conquer still further the adverse environments with which it has to contend. We have in these Desert regions the two main conditions for the evolution of species according to received doctrines : (1) Abundance of unoccupied space, and (2) Recent change in conditions of climate, involving variation and a struggle for existence. Besides an extreme dryness of climate little suited to vegetable growth, the vast numbers of rodents in the Desert must operate largely to check an increase of plant-life; and it is in accordance with modern views to suppose that the excessively obnoxious qualities of smell, viscidity and prickliness so commonly met with may here be of real protective value to their owners. A species which jerboas and sand-rats found preferable to all others would speedily disappear. Those forms which are 'in harmony with their environment' have, however, abundant room to make use of, and it is often a matter of surprise a particular spot is selected by a plant when areas with precisely similar conditions which might be similarly occupied are devoid of growth.

With regard to the qualities above alluded to I may be here allowed a few words, although this subject has no doubt been often gone into. Dr. Anderson, in his excellent 'Florula Adenensis,' already referred to, has left little to be told. The more numerous flora of Sinai gives, however, wider groups of examples. One speedily comes to the conclusion that a Desert plant will be either (1) *Hoary* or *white*, with *woolliness*, as Morettia, Glinus, Farsetia, Reaumuria, Tribulus, Dianthus, Lotononis, Neurada, Anvillæa, Odontospermum, Pulicaria, Verbascum, Stachys, Teucrium, Ærua; (2) *Glaucous*, as Moricandia, Zilla, Nitraria, Capparis, Microrhynchus, Pyrethrum; (3) *Viscid*, or *gummy*, as Cleome, Nepeta, Pulicaria, Acacia, Alhagi, Hyoscyamus; (4) *Hooking*, or *sticking with hairs*, as Trichodesma, Cleome, Andrachne, Forskahlea ; (5) *Prickly* or *spiny*, as Zilla, Capparis, Nitraria, Acacia, Alhagi, Astragalus, Microrhynchus, Sporobolus, Iphiona ; (6) *Strongly aromatic*, as Deverra, Pyrethrum, Achillea, Varthamia, and many labiates ; (7) *Nauseating*, as Peganum, Haplophyllum (Ruta), Cleome, Odontospermum ; (8) *Whitened, as if calcified* (usually stems), as Scrophularia (edges of leaves), Astragalus, Microrhynchus, Zollikoferia, Calligonum, Ephedra, Tamarix, Salsola ; (9) *Succulent*, as Nitraria, Salsolæ, Zilla, Moricandia, Boucerosia, Reaumuria. These qualities are often combined in various degrees, and all the sets of examples might be easily increased. They seem to be the outcome of extreme heat and dryness; the latter reduces the tissues and develops spines, and appears also to be favourable to the growth of pubescence in plants. Heat promotes the growth of glandular and odoriferous essences, which a dry atmosphere will intensify; while, no doubt, many of these modifications are developed, or at least intensified in accordance with the requirements of the species for purposes of protection from enemies, or perpetuation of kind.

This flora, with these strongly-marked features, extends without interruption from the Cape de Verde Islands and Senegal, across the African and Libyan deserts to Egypt and Nubia, thence across Arabia, the southern part of Mesopotamia and Southern Persia, throughout Beloochistan, South Afghanistan, Scinde, and into the Punjaub. It thus ranges along a belt of surface upwards of 5,000 miles in length from east to west, thinning out at the eastern and western extremities, and widest in the African Desert, where it extends over the space between the tenth and thirty-seventh parallels of latitude. Continuity of representative or identical flowering plants for so wide a range is without a parallel elsewhere on the earth's surface. The nearest approach to a parallel lies in the circumpolar regions. In many cases the species is identical throughout; in others closely allied forms represent one another in different countries along the belt. A few conspicuous plants may be mentioned, which extend from the Cape de Verdes and Canaries to India: Citrullus colocynthis, Aizoon canariense, Francoeuria crispa, Picridium tingitanum, Trichodesma africanum, Suæda vermiculata, Ærua javanica, Euphorbia granulata, Andropogon foveolatus and Aristida cærulescens. This flora occupies, in fact, most of the area which was in Tertiary times a sea-bottom extending from the Bay of Bengal to the Atlantic.[1]

On looking to the second column, where these widespread Desert species are thrown together, an illustration of the central position of Sinai with regard to this flora ranging so far east and west will be found. There are 177 Desert species, with a wide range east, west, or south, or in any two, or in all of these directions. Of these, 112 extend far eastwards, 105 extend far westwards, and sixty have both directions; so that

---

[1] See Wallace's 'Geog. Dist.,' i., p. 286.

Sinai, standing somewhat nearer to the eastern extremity, has a very slight preponderance of eastern Desert forms.

Uniformity such as this would tend to induce the belief that the conditions had been long unaltered. But geology[1] teaches us that during the glacial epoch much of the Sahara was a sea-bed, and with glaciers on Lebanon, and probably perpetual snow on Mount Sinai, our flora, in whatever form or quantity it existed, must have at that time held more southern ground. Until we are better acquainted with the features of the interior of Southern Arabia, it is premature to indulge in further speculation.

MEDITERRANEAN FLORA OF SINAI.

The next column, that of the Mediterranean species properly so called, is chiefly remarkable for its limited numbers. In the widest sense in which this term is used, this list would be increased by the addition of several from my Plateaux and Desert groups. But of typical Mediterranean species very few can stand the droughts of Sinai, or else have become so modified thereby that they now rank as distinct forms. The number of non-European genera already enumerated best illustrates the *un*-Mediterranean character of the flora. The proportion of this group to the total is about one in six; to the Desert total about one in four. Here, as in the last column, we are confronted with a majority of eastern plants.

PLATEAUX OR MONTANE FLORA OF SINAI, AND GENERAL REMARKS.

The Plateaux flora of Sinai is in some respects the most interesting. The flora plants of the highest points are sometimes widespread Mediterranean species, as Herniaria, Capparis, Poterium, and others; but the majority are plants of the Persian and Syrian elevated regions, or of those lying due north in Asia Minor and in the Taurus region. Of the alpine Lebanon species very few occur; a Caryophyllaceous species (Gypsophila hirsuta, δ), an Astragal (A. echinus), and a Campanula (C. dulcis), are the only representatives on Sinai. Gypsophila and an Arenaria (A.

---

[1] See Lyell, 'Principles of Geology,' vol. i., p. 253; Wallace, 'Distribution of Animals,' vol. i., p. 39.

graveolens) may be quoted as the only high-mountain plants of South-eastern Europe which have found their way to Sinai. From the Nubian and Abyssinian mountains we have Colutea, Gaillonia, Otostegia, and a Micromeria; while from east and north-east, chiefly in Persia, we meet with such genera as Roemeria, Deverra, Zozimia, Arnebia, Alkanna, Echinospermum, and Paracaryum, and a number of showy labiates and scrophulariaceous species which often have here their western limit.

The Cryptogamic vegetation of Sinai is of a very widespread description, as is usually the case with these forms of life. The ferns maidenhair, Cheilanthus, and Nothochlœna illustrate this, the latter two being included in the list of Sinai plants on the authority of Oliver's appendix to the Ordnance Survey. The maidenhair every traveller has observed. I did not myself meet the other two ferns until reaching Mount Hor, which lies outside the peninsula. The Sinaitic mosses, with one exception, are all British. I refer the reader back to my notes on the spot for further information on this subject.

In order to obtain a fuller insight into the character of the Sinaitic flora it is necessary to introduce a few geological remarks.

It is rendered probable by many considerations, as Professor Hull has pointed out, that in recent times a far wetter climate prevailed in Sinai—a condition of things which was probably continuous throughout the glacial period and subsequently. Without evidence on the spot, this might have been deemed probable; but such evidence is forthcoming, and I refer the reader to Professor Hull's work for fuller information.

In this period there were large lakes in Sinai; several of these dried-up basins are very noticeable. Again, during some portion of Pliocene times, and perhaps down to the earliest part of the glacial epoch, Sinai was probably dissevered from Africa, and form?d part of the Euro-Asiatic continent. Evidence derived from raised beaches and their contents on the shores of the Red Sea and the Mediterranean in several places has, I believe, established this. I would refer the reader here to Issel's ' Malacologia del Mar Rosso,' who draws this conclusion from a comparison of the present and sub-fossil shells of the Mediterranean and Red Seas. Thus, whatever flora (probably a more temperate one than the present) the northern parts of Africa supported were cut off at that period from entering Sinai by a direct route.

PLATEAUX OR MONTANE FLORA.

During the colder part of the glacial period, when Lebanon maintained glaciers to about 4,000 feet below the summit, and the Jordan Valley was most likely a series of lake-basins, all, including the Dead Sea, much more extensive than at present, the climatic condition of Sinai must have been about equivalent to that of the British Isles at the present period, with an abundant rainfall, and the then alpine flora of Lebanon may perhaps have extended to Sinai. It is reasonable to infer that Lebanon has had a more plentiful alpine flora than the very few existing remnants would seem to imply. Mountains in the Himalayan region at similar altitudes and further south are far better stocked with Arctic species. Sir Charles Lyell, commenting on this southern extension of the glacial severity to Lebanon, first noticed by Sir Joseph Hooker, thinks it probable that it is referable to the earliest portion of the 'great ice age,' in which case the Arctic species that must have once decked the upper heights of Lebanon have had all the greater time to perish. Although at present the snow never quite leaves the summits of Lebanon and Hermon, the summer atmosphere is drier and warmer than an alpine flora would thrive under; on the barren Sinaitic mountains this is, of course, still further the case.

At this period no doubt Sinai was abundantly vegetated. The more temperate northern and north-eastern floras were pushed southwards by the northern cold, their predecessors being either exterminated, so far as Sinai is concerned, or, if allowed sufficient time, highly modified or driven south in their turn along the adjoining continents. As milder influences began to prevail more and more, Arabian forms crept in; the connection of Sinai with that country across the Arabah watershed remaining unsevered. Finally, when the glacial period mollified, and the temperate forms retired upwards, the Arabian flora steadily advanced northwards from the south of its peninsula, and overran the lower parts of Sinai. With the elevation of the Sinaitic peninsula, and a gradual more or less lagoon-divided connection with Africa, corresponded an increased temperature, and Nubian and other African forms also spread northwards, and finally eastwards. At the same period, and subsequently, a great portion of the present flora of the Delta obtained its opportunity for extending westwards, and hence the number of Arabian forms which occur in Lower Egypt. This elevation must have continued until it carried the bottom of the shallower parts of the Red Sea above the surface, especially of those two arms which wash the sides of Sinai. The uniformity of the lowland floras on the opposite side of the Red Sea drives me to the conclusion that there was at some period much more continuous land for the migration of species between Africa and Arabia than there is at present, and in consequence of this community of species being most observable amongst the so-called Desert forms, which are in many cases, perhaps, of recent origin, and form, in fact, the existing characteristic life, and because the older flora of more temperate times is much more distinct on both sides of the Red Sea, I imagine that this continuity of surface belongs to the more recent period. In speaking of this community of flora, it is necessary to point out that a considerable portion, perhaps a third part, of the Sinaitic Desert flora may be considered as belonging specially to the neighbourhood of Sinai, almost confined, in fact, to Arabia Petræa (which includes Sinai and South Palestine) and Egypt. So that it seems fair to regard Sinai, or perhaps more properly the Delta, as a centre of dispersion for a considerable number of Desert forms. It is usually conceded that, at the close of the glacial period, the temperature kept on increasing to a greater point of warmth than has since prevailed. With sufficient atmospheric moisture, and sheets of fresh water gradually diminishing, no doubt, with an increasing heat, Sinai, open to southern visitors both from Africa and Arabia, must have been well stocked with what we now consider tropical or sub-tropical forms. In the Wâdy 'Arabah, for example, two large saline lakes, at least, existed, and its sides were probably fringed throughout with such woods as now exist in a reduced condition only in the Ghôr es Safieh. In the drier parts the progenitors or ancestral types of the present highly-modified Desert forms spread rapidly. When I come to analyze the flora of the Ghôr es Safieh, these remarks will obtain greater force. It will be seen that the majority of tropical forms surviving there on account of its peculiar condition are Nubian, and have probably arrived thither by way of the 'Arabah.

To illustrate the Arabian—for we may presume that most of the Persian Sinaitic plants are also Arabian—character of the Sinaitic Plateaux flora, I may mention that the following species, believed peculiar to Sinai, were found by me on Mount Hor on the western skirts of the great Arabian tableland: Pterocephalus sanctus, Varthamia montana, Echinops glaberrimus, Phlomis aurea, and Teucrium sinaicum, all of which are Plateaux or Montane species; while Crucianella membranacea, Picris cyanocarpa, and the form Mentha lavandulacea, also believed peculiarly Sinaitic, have recently been discovered in Midian by Captain Burton.

At the close of the cold period, which favoured the dispersion of these species (and no doubt the formation of some of them), a warm post-glacial time intervened, and gave the majority of the tropical plants of the Desert oases (as of the Ghôr and Wâdy Feirân) the means of spreading throughout their present wide range, while at the same time it committed ravages on whatever northern plants had reached Sinai during the ice age. The dry parched forms which care not where or how they exist, so long as it is a hot desert, need no further means of dispersal than a sandstorm, which will whisk their seed-laden stems to inconceivable distances; these, which are really the *bonâ-fide* inhabitants of the Desert, are easily distributed by the means now existing. There are Desert-birds who follow them, and pick their seeds from place to place with as wide a range often as the plants themselves. Many of these plants, again, as has been mentioned, are extremely sticky, and furnished with hooked hairs, which attach them to the skins of beasts who have to travel far for food. And there are the mighty and dreaded storms of these wildernesses, sufficient in themselves, where the whole surface shifts, to spread identical vegetation from end to end of the sandy or dusty sea. Thus the time allowed for the wide spread of these species will not, perhaps, in view of the many favourable means, appear too scanty. In connection with this it would be necessary to study how far the seeds of these Desert species which reach the Canaries and Cape de Verdes are fit for transmission by the recognised sources by which islands may be stocked.[1]

The present condition of Sinai, that of extreme drought, has no doubt been intensified by the consumption of timber in historic times. The Israelitish hosts and the Egyptian mining companies must have all contributed to this unfortunate result, and the charcoal-burning Bedouins are still engaged in carrying out the work of destruction upon the few remaining groves of acacia, nubk, and tamarisk in Sinai.

[1] See Wallace's 'Island Life,' pp. 248-252.

AN ANALYSIS OF THE FLORA OF SINAI.

In order to give a full idea of the relative distribution of the Sinaitic flora over the world, I have compiled the following table. It must be borne in mind, however, that this table gives equal value to all forms, several of which can hardly be deemed worthy of specific rank.

The total number in my list is 509; of these seven species (414 to 420) are omitted, being of marine distribution, and fourteen other cosmopolitan weeds may also be left out. These latter are chiefly weeds of cultivation, and the number so dealt with might fairly be increased. I have omitted Nasturtium officinale, Sisymbrium irio, Sinapis arvensis, Malva rotundifolia, Portulaca oleracea, Convolvulus arvensis, Sonchus oleraceus, Solanum nigrum, Chenopodium album, C. murale, Juncus bufonius, Scirpus maritimus, Cynodon dactylon, and Eleusine indica. The number of Sinaitic plants whose distribution is treated of is, therefore, 488. The localities are mainly from Boissier's 'Flora Orientalis.' My Midian authority is Oliver's appendix to Captain Burton's 'Land of Midian,' and Balfour's appendix to the same author's 'Gold-mines of Midian.' For Aden I have used Dr. Anderson's 'Florula Adenensis.' The Cape de Verde list has been increased, as also has Madeira, from the compilation in Hooker's 'Flora Nigritiana.' For Palestine, Canon Tristram's list is relied on; but, as I have already stated, Monsieur Boissier is my main authority, Nyman's 'Conspectus' being referred to when necessary.

Species confined to Sinai	33	
Species confined to Sinai and Arabia Petræa, including southern borders of Palestine	30	63
Sinaitic species found in Palestine[1]		387
,, ,, Egypt		293
,, ,, Persia		187
,, ,, North Africa	162	221
,, ,, interior of North Africa	59	
,, ,, Europe	137	
Sinaitic species found in local, arid or desert European stations, chiefly at Sicily, Cyprus, Crete or Southern Spain	21	158
Sinaitic species found in Afghanistan and Beloochistan		128
,, ,, Syria		139

[1] The total Sinaitic species ranging beyond Arabia Petræa and South Palestine, here dealt with, is therefore 425, 63 being endemic.

PLATEAUX OR MONTANE FLORA.

Sinaitic species found in Asia Minor	113
,, ,, Midian	120
Sinaitic species found in Arabia, excluding Aden, Muscat, Midian and Arabia Petræa (=Arabia Trop., Arabia Felix, Arabia Deserta)	79
Sinaitic species found east of Arabia, in Persia, India, Afghanistan, Beloochistan and Mesopotamia	240
Sinaitic species found in Aden (S.W. Arabia)	37
,, ,, Muscat (S.E. Arabia)	37
Sinaitic species found in all Arabia, excluding Arabia Petræa, to which Sinai belongs	191
Sinaitic species found in Mesopotamia	93
,, ,, India	74
Sinaitic species found in east of Arabia Petræa, either in Arabia, Mesopotamia, Afghanistan, Beloochistan or India	328
Sinaitic species found in Turkestan, Songaria, and countries around the Caspian	65
Sinaitic species found in Nubia	73
,, ,, Abyssinia	53
Sinaitic species found in Senegal, Senegambia and tropical Africa	25
Sinaitic species found in Cape de Verde Islands	36
,, ,, Madeira	29
,, ,, Canaries	73
,, ,, Cape of Good Hope	19
,, ,, Azores	6

Although I am averse to drawing conclusions from lists of figures, I would point out that, considering how little we know of the botany of Central Arabia, two of the above totals are interesting. Of the 425 Sinaitic species ranging beyond Arabia Petræa, 191 are known to inhabit Arabia at different parts, whether it be Aden, Midian, Muscat, Arabia Felix, Yemen, or tropical Arabia. But of the same Sinaitic total there are no less than 240 found east of Arabia in India, Persia, Afghanistan, Beloochistan, or Mesopotamia. No doubt many of those fifty plants not recorded from the intervening country do not skip the Arabian peninsula.

If we omit the adjoining countries, Arabia on the east and south-east, and Egypt, Nubia, and Abyssinia on the west and south-west, and compare the numbers of plants common to Sinai, and those countries east of Arabia on the one side, and west of Eastern Africa—that is to say, in the

156   AN ANALYSIS OF THE FLORA OF SINAI.

north and north-west of Africa—with the Atlantic islands, on the other side, we come on an unexpected result, the two totals being almost exactly equal. The increased number of Mediterranean species westwards almost exactly counterbalances the majority of eastern forms. Several of these western forms are weeds of cultivation.

In concluding these remarks, I will observe that an analysis of the distribution of the nearest allies to the endemic species of Sinai does not lead to any distinct result, these being sometimes Desert and sometimes Plateaux. A good proportion, however, point to eastern and north-eastern sources for the original home of the Sinai forms.

ON THE FLORA OF THE GHÔR, OR VALLEY OF THE DEAD SEA.

I have already mentioned that I gathered about 225 sorts of flowering plants at the south end of the Dead Sea. Many of these are not found in Canon Tristram's list, which contains, on the other hand, several I did not meet with. Again, at the northern end of the Dead Sea, in the neighbourhood of Jericho, a good many more species occurred, enough to raise the total to at least 250. Here, and in other oases around this barren sheet of water, Canon Tristram has found different southern species, chiefly at Engedi and 'Callirhoe.' This group, enlarged by my additions, may be regarded as an arm of the Desert flora thrust up the Wâdy 'Arabah to its northern limit. The remainder of the Ghôr plants are chiefly Mediterranean species, often of a very local type, or ranging east to Persia. Many widespread European species also occur, especially at Jericho.

The past climatic conditions of Sinai have been already spoken of. The tropical flora of the Ghôr (which I will presently enumerate) probably dates from a time when Sinai had a richer flora than at present, and when a continuous vegetation, similar to that at present in the Ghôr es Safieh, but probably of largely increased variety, extended from Nubia and Arabia, most prominently, no doubt, along the valley of the 'Arabah.

Since this period the chain has been broken in various places, and many forms now isolated in the Ghôr have their nearest habitat in Sinai, while not a few are found no nearer than Nubia, Abyssinia, and tropical

Arabia. This local extermination is due almost entirely to drought. There may or may not have been a warmer period since the glacial, but with increased moisture and the present temperature the lower valleys of Sinai and the 'Arabah would be quite capable of supporting all the hotter plant-life of the Ghôr.

Of the present 'Desert flora' of the Ghôr—I mean that which is properly so called—there is little to say. These are species which require no oases to exist in, and have no difficulty in spreading. It is a portion of this group of plants that is dealt with by Mr. Lowne in his excellent essay on the floras of the Ghuweirah and Mahauwat Wâdies at the southwest end of the Dead Sea.

Canon Tristram, in dealing with this subject, travels back to miocene times for a solution of the existence of Nubian and Ethiopian forms in the Ghôr. With all deference, I regret to find myself unable to agree with him. Were his theory confined in its conclusions to Palestine—a country with whose natural history no one is so familiar—I should hardly venture to express an opinion; but when he derives this flora from miocene times, we have Professor Forbes' old theory served up afresh, and the advent of our so-called 'Spanish flora' in Ireland is made synchronous with that of the Ghôr from its more southern home. Forbes' theory dates the accession of the Spanish group of plants still lingering in Kerry and Connemara, but at no intermediate points, to ante-glacial times, when, it was suggested, there may have been continuous or contiguous land from Kerry to Spain. But geologists have to meet facts, and I think no botanist will now admit that these remnants of a southern flora, able, and no more, to hold their own in the present mild Irish climate, could have existed throughout the glacial period, when, in all probability, the south-west of Ireland was submerged in a frozen sea for several hundred feet below its present level. As I have always had a pet aversion to this part of Professor Forbes' theory (which I have seen recently more than once quoted), so I feel compelled to disagree with Canon Tristram, who looks on the tropical flora of the Ghôr as a 'northern outline' dating from miocene times. I have already alluded to the probable climatic condition of the Jordan Valley during the glacial epoch. It seems to me quite impossible that the more tropical portion of its now-existing life could have then survived, supposing it to have arrived there previously. It would be as reasonable to expect that the osher, the salvadora, and the false balm of Gilead would now survive if they were planted out in the south of England. The Irish question alluded to above is difficult of solution—Irish questions usually are—and whether we are helped by an appeal to the Holy Land may be doubted.

Most of the species here alluded to may be readily found on reference to my Sinai list, where the majority occur, and those plants which are there marked as tropical, and occurring in Palestine, almost invariably occur in the Ghôr. A few others, however, have to be added which do not occur in Sinai. In the following list I will put together the most remarkable tropical species. They belong almost entirely to the Ethiopian region of Wallace, as will be seen from their appended geographical distribution. The list might easily be increased by the addition of several species found in Nubia and Abyssinia, but which range northwards in other parts, as in Persia or Africa, and whose headquarters may be regarded as doubtful. I prefer to call especial attention to those which attain here an unexpectedly far northern limit. It will be seen that they are, almost without an exception, natives of Nubia. With reference to Midian plants, I shall give a brief analysis of Burton's list. Of 163 species, all except forty-one occur in Sinai, and of these forty-one, twenty-six occur in Palestine (nine in the Ghôr), and fifteen elsewhere. This illustrates the fact that the Ghôr has received Nubian species either by way of Aabia or Sinai and Egypt, for most of these Midianitic non-Sinaitic species are also Nubian. It serves also to show the community of species on opposite sides of the Red Sea. The remaining fifteen Midian plants not found in Sinai or Palestine are chiefly tropical, or more southern than Sinai, with endemic Arabian and a couple of north-eastern Plateaux species.

TROPICAL FLORA OF THE DEAD SEA BASIN.

MENISPERMACEÆ.

Cocculus Leæba, D.C. A native of Cape de Verdes, Senegal, Senegambia, Nubia, Upper Egypt and Abyssinia, occurring in Sinai and the Wâdy 'Arabah to the Dead Sea.

PARONYCHIEÆ.

Sclerocephalus arabicus, Boiss. Cape de Verdes, interior of North Africa, extreme south of Persia and Muscat in South Arabia. Sinai and Ghôr es Safieh. A desert species.

MALVACEÆ.

Abutilon fruticosum, G. et P. Senegal, Nubia, Abyssinia, Arabia (Aden and Midian), Scinde and Beloochistan. Sinai and in Edom, along the 'Arabah to the Dead Sea.

A. muticum, Del. Cape de Verdes (Webb), Senegal, Nubia, Upper Egypt, Arabia; Southern Persia, Afghanistan. Sinai and Ghôr es Safieh.

TILIACEÆ.

Corchorus trilocularis, Linn. Tropical Africa and Asia. Cape de Verdes. Ghôr es Safieh. Not found in Sinai.

CAPPARIDEÆ.

Capparis Sodada, R. Br. Upper Egypt, Nubia, Abyssinia, Southern Persia and North-west India. Not found in Sinai, but occurs in Midian. North-east end of the Dead Sea.

SIMARUBEÆ.

Balanites ægyptiaca, Del. Upper Egypt, Nubia, Abyssinia and Arabia. Not found in Sinai. Ghôr es Safieh and Jericho.

AN ANALYSIS OF THE FLORA OF SINAI.

MORINGEÆ.

Moringa aptera, Gært. Upper Egypt, Nubia, Muscat and tropical Arabia. Sinai and Ghôr.

LEGUMINOSÆ.

Rhynchosia minima, D.C. Tropics of both hemispheres and Cape de Verdes. Not found in Sinai. Ghôr es Safieh.

Indigofera paucifolia, Del. Upper Egypt, Senegal, Nubia and Eastern India. Midian in Western Arabia by the Red Sea. Ghôr es Safieh.

Cassia obovata, Collad. Cape de Verdes and Senegal, Nubia, Abyssinia, Upper Egypt; Southern Persia and Beloochistan; Aden, Midian and Arabia Felix, Scinde, 'Jamaica and Texas.' Sinai and Ghôr es Safieh.

Prosopis spicigera, Linn. Scinde and tropical India; Southern Persia and Beloochistan; Muscat in South-east Arabia. Ghôr.

Acacia Seyal, Del. Upper Egypt, Senegal, Nubia, Abyssinia, Muscat and Arabia Felix. Sinai and along the 'Arabah to the Ghôr.

A. læta, R. Br. Upper Egypt, Nubia, Abyssinia and Arabia Felix. Ghôr es Safieh. Not found elsewhere north of the Soudan.

CUCURBITACEÆ.

Cucumis prophetarum, Linn. Upper Egypt, Nubia; Muscat, Midian, Aden and Arabia Felix; Scinde. Sinai and the Ghôr.

FICOIDEÆ.

Trianthema pentandrum, Linn. Senegal, Nubia and Abyssinia; Midian and tropical Arabia; Southern Persia, Afghanistan and Northern India. Not found in Sinai. Ghôr es Safieh.

COMPOSITÆ.

Erigeron Bovei, D.C. (sub *Conyzâ*). Upper Egypt (?); Muscat and Midian in Arabia; Sinai and Ghôr es Safieh.

TROPICAL FLORA OF THE DEAD SEA BASIN.   161

Pulicaria undulata, Linn. (sub *Inula*). Middle and Upper Egypt, Nubia and Senegal ; Midian. Sinai and Ghôr es Safieh.

N.B.—Several other Nubian and Abyssinian Compositæ occur, but they are usually widespread in Egypt and North Africa.

SALVADORACEÆ.

Salvadora persica, Gærtn. Middle and Upper Egypt, Nubia, Abyssinia, interior of Algeria and Central Africa ; Midian, Aden and tropical Arabia ; North-west and tropical India. Sinai and Ghôr es Safieh.

ASCEPIADEÆ.

Solenostoma argel, Del. Upper Egypt and Nubia ; Sinai and Ghôr.

Calotropis procera, Willd. Upper Egypt, Nubia, Abyssinia, subtropical Algeria, and Senegal ; Southern Persia and Afghanistan ; Northern India, Midian. Sinai and Ghôr.

Pentatropis spiralis, Forsk. Senegal, Nubia and Abyssinia ; Muscat and tropical Arabia. Wâdy 'Arabah and Ghôr es Safieh.

Glossonema Boveanum, Dcne. Upper Egypt, Nubia and Abyssinia ; tropical Arabia. Eastern Ghôr. Not found in Sinai.

Leptadenia pyrotechnica, Forsk. (sub *Cynancho*). Upper Egypt, Nubia and Senegal ; Southern Persia. Sinai and Ghôr es Safieh.

N.B.—*Dæmia cordata* and *Oxystelma alpina*, Nubian species, also occur, but they are found in Lower Egypt.

BORAGINEÆ.

Heliotropium arbaniense, Fres. Upper Egypt and Nubia ; Midian and tropical Arabia ; Sinai and Ghôr es Safieh.

Trichodesma africanum, R. Br. Cape de Verdes, Senegal and Upper Egypt; Beloochistan and Northern India ; Midian and Muscat in Arabia. Sinai and up the 'Arabah to Ghôr es Safieh.

SOLANACEÆ.

Solanum coagulans, Forsk. Upper Egypt, Nubia and Abyssinia; Southern Persia and Beloochistan; Midian and tropical Arabia. Not found in Sinai. Ghôr.

Lycium arabicum, Sch. Upper Egypt, Nubia; tropical Arabia; Northern India. Sinai and in the 'Arabah to the Ghôr.

SALSOLACEÆ.

Atriplex farinosum, Forsk. Upper Egypt and tropical Arabia. Sinai and along the 'Arabah to the Ghôr.

Suæda monoica, Forsk. Upper Egypt, Nubia and Abyssinia; Midian and tropical Arabia; India. Sinai, 'Arabah, and Ghôr es Safieh.

Salsola fœtida, Del. Middle and Upper Egypt, Nubia, Senegal; tropical Arabia; Beloochistan and Northern India. Sinai, 'Arabah, and Ghôr es Safieh.

AMARANTACEÆ.

Digera arvensis, Forsk. Nubia and Abyssinia; Arabia; Afghanistan and India. Ghôr es Safieh. Not found in Sinai.

NYCTAGINEÆ.

Boerhavia verticillata, Poir. Nubia, Abyssinia, and Senegal; Muscat in South-east Arabia. Wâdy 'Arabah and Ghôr es Safieh. (Also near Gaza.)

LORANTHACEÆ.

Loranthus acaciæ, Zucc. Nubia, Abyssinia and Wâdy 'Arabah to the Ghôr.

EUPHORBIACEÆ.

Euphorbia ægyptiaca, Boiss. Middle and Upper Egypt, Nubia, Abyssinia, tropical Africa, Senegal, Cape de Verdes; Northern India; Ghôr es Safieh. *Euphorbia granulata, Crozophora obliqua*, and *Andrachna aspera* might also be quoted, but they have other outlying northern habitats.

ADDITIONS TO FLORA OF PALESTINE.

CYPERACEÆ.

Cyperus elcusinoides, Kunth. Afghanistan; tropical Africa; India and Australia. Ghôr es Safieh.

N.B.—Several tropical grasses occur in the Ghôr, but, like the majority of the Compositæ, they are too widespread to be illustrative.

The following is a list of species obtained by me in Palestine (including Petra and Mount Hor), which are additions to its flora as given by Canon Tristram. Except where otherwise stated, the species have been determined by Mr. Oliver.

ADDITIONS TO PALESTINE FLORA.

MENISPERMACEÆ.

Cocculus Leæba, D.C. Var. approaching *C. villosus*. Ghôr el Feifeh and es Safieh.

CRUCIFERÆ.

Matthiola humilis, D.C. Tell Abu Hareireh to Gaza. Known only from Egypt in one or two places. Determined by Mons. Boissier.

Sisymbrium erysimoides, Desf. Ghôr es Safieh and Jericho. A widespread desert species.

Enarthrocarpus lyratus, D.C. Bîr es Sebâ to Gaza; Jericho. Known from Greece, Lower and Middle Egypt.

RESEDACEÆ.

Caylusea canescens, St. Hil. In Wâdy 'Arabah near the Dead Sea, and between the Ghôr and Bîr es Sebâ. A desert species.

SILENEÆ.

Silene Hussoni, Boiss. Between Bîr es Sebâ and Tell Abu Hareireh. Named by Mons. Boissier, who has described it from its only other locality, 'Wâdy Sannour deserti Ægyptiaco Arabici.'

S. colorata, Poir. Jericho. Not found east of Crete hitherto. Not in Boissier's 'Flora Orientalis.' See Nyman's 'Conspectus.'

PARONYCHIEÆ.

Polycarpon succulentum, Del. At Jericho and between Tell Abu Hareireh and Gaza.

Paronychia nivea, D.C. Bir es Sebâ. Not in Boissier's 'Flora Orientalis.' Found in Greece (Nyman), but no further east.

P. desertorum, Boiss. Wâdy Ghuweir, between Petra and the Ghôr in Edom. A desert species.

Sclerocephalus arabicus, Boiss. Ghôr es Safieh. A desert species.

TAMARISCINEÆ.

Tamarix articulata, Vahl. Between Ramleh and Jerusalem, introduced (?).

ZYGOPHYLLEÆ.

Zygophyllum simplex, Linn. Ghôr es Safieh.

LEGUMINOSÆ.

Ononis campestris, Koch. Gaza.

Indigofera paucifolia, Del. Ghôr es Safieh. Known from Upper Egypt and Midian.

Colutea aleppica, Lam. Summit of Mount Hor.

Astragalus acinaciferus, Boiss. Wâdy 'Arabah immediately above the Ghôr.

Rhynchosia minima, D.C. Ghôr es Safieh. A common tropical species.

Acacia læta, Br. Ghôr es Safieh. Also tropical.

FICOIDEÆ.

Trianthema pentandra, Linn. Ghôr es Safieh.

RUBIACEÆ.

Rubia peregrina, Linn. (?) Petra and rocky slopes of Mount Hor above Petra.

Galium petræ, n. sp. Petra.

DIPSACEÆ.

Pterocephalus sanctus, Dcne. Petra and Mount Hor.

COMPOSITÆ.

Varthamia montana, Vahl. Mount Hor.
Erygeron (Conyza) Bovei, D.C. Ghôr es Safieh.
Eclipta alba, Linn. Ghôr es Safieh.
Tripteris Vaillantii, Dcne. Petra and Mount Hor.
Echinops glaberrimus, D.C. Wâdy Harûn.
Scorzonera alexandrina, Boiss. Tell Abu Hareireh to Gaza. Determined by Mons. Boissier. Known from Tunis, Algiers, and Alexandria.
Sonchus maritimus, Linn. Ghôr es Safieh by the edge of the Dead Sea.
Zollikoferia Casinianæ, Jaub. (?) Mount Hor and Ghôr es Safieh, but specimens incomplete. Known from Egypt and shores of Red Sea.
Z. stenocephala, Boiss. (?) vel *Z. arabica*, Boiss. (?) Specimens incomplete. Wâdy Zuweirah.
Crepis senecioides, Del. Bethlehem, near Solomon's Pools. Known from Cairo, Alexandria and Sackara in Egypt.

ASCEPIADEÆ.

Pentatropus spiralis, Forsk. Ghôr es Safieh and Wâdy Harûn. Not in flower. A tropical species.
Boucerosia Aaronis, n. sp. Wâdy Harûn and Mount Hor.

GENTIANEÆ.

Erythræa spicata, Pers. Ghôr es Safieh.

SCROPHULARIACEÆ.

Celsia parviflora, Dcne. Ghôr es Safieh.
Linaria macilenta, Dcne. Wâdy Harûn.

Scrophularia heterophylla, Willd., *forma*. Petra and Mount Hor. Known only from Greece and its confines (?).

Lindenbergia sinaica, Dcne. Wâdy Ghuweir and Ghôr es Safieh.

LABIATÆ.

Micromeria sinaica, Bth. (?) Rocky ledge above the Tufileh River near the Ghôr.

Salvia deserti, Dcne. Wâdy Harûn.

Phlomis aurea, Dcne. Mount Hor.

Teucrium sinaicum, Boiss. Mount Hor and Petra.

PLANTAGINEÆ.

Plantago Loefflingii, Linn. Jericho.

SALSOLACEÆ.

Atriplex alexandrinus, Boiss. Jebel Usdum; Jericho. Known only from Tunis and Alexandria. Determined by Mons. Boissier.

A. leucocladum, Boiss. Ghôr es Safieh. Determined by Mons. Boissier.

Salsola inermis, Forsk. Jebel Usdum; Petra; Gaza. Known only from Forskahl's locality, near Alexandria. Determined by Mons. Boissier.

S. longifolia, Forsk. Jebel Usdum. Known only from Egypt.

S. fœtida, Del. Jebel Usdum.

Anabasis setifera, Moq. Jebel Usdum; Tell Abu Hareireh to Gaza.

AMARANTACEÆ.

Digera arvensis, Forsk. Ghôr es Safieh and Ghôr el Feifeh. A tropical species.

NYCTAGINEÆ.

Boerhavia verticillata, Poir. Wâdy Harûn and Ghôr es Safieh. Between Gaza and Tell Abu Hareireh. A tropical species.

B. repens, Linn. Ghôr es Safieh and Jericho. Also tropical.

ADDITIONS TO FLORA OF PALESTINE.

THYMELÆACEÆ.
Daphne linearifolia, sp. nov. Petra.

EUPHORBIACEÆ.
Euphorbia ægyptiaca, Boiss. Ghôr es Safieh. A tropical species.

SALICINEÆ.
Salix acmophylla, Boiss. Wâdy Ghuweir and Ghôr es Safieh.

TYPHACEÆ.
Typha angustata, B. et C. Ghôr es Safieh?

LILIACEÆ.
Urginea undulata, Desf. Wâdy 'Arabah, near the Ghôr, and at Bîr es Sebâ.

CYPERACEÆ.
Cyperus lævigatus, Linn. Gaza.
C. eleusinoides, Kunth. Ghôr es Safieh. A tropical species.

GRAMINEÆ.
Panicum molle (*P. barbinode,* Trin.). Jericho. Not included in 'Flora Orientalis.'
Pennisetum dichotomum, Forsk. Wâdy 'Arabah, near the Ghôr.
Sporobolus spicatus, Vahl. Near Gaza?
Agrostis verticillata, Vill. Jericho and near Bethlehem.
Danthonia Forskahlia, Vahl. Wâdy 'Arabah, near the Ghôr.
Eragrostis poæoides, P. de B. Ghôr es Safieh.
E. pilosa, Linn. Ghôr es Safieh.
E. megastachya, Link. Ghôr es Safieh.

EQUISETACEÆ.
Equisetum elongatum, Willd. Ghôr es Safieh and Gaza.

CHARACEÆ.

Chara hispida, Linn. 'Ayûn Buweirdeh, in the 'Arabah.

MUSCI.

In this group Mr. Oliver has kindly obtained for me the assistance of Mr. William Mitten, A.L.S.

Mosses were very sparingly met with in Sinai, and only at great elevations. A few occurred on Jebel Mûsa, from about 6,500 feet to the summit (7,385 feet), under the shady rocks in a gully looking north. At about the same height (6,500 feet) on Jebel Katharîna there is a spring, Mayau esh Shunnar (' Fountain of the partridge '), where several patches were obtained, and a few more up the valley from here to the summit, 8,536 feet. No others were gathered in Sinai, nor until entering the country of Edom on the east side of the 'Arabah at Wâdy Abu Kosheibeh (' W. Harûn '), and high up Mount Hor. Others were found in Palestine at their cited localities. It is most probable that the Mount Hor district would yield several more forms of musci, especially if visited about one month later in the year than my hurried day there on December 9. Almost immediately afterwards (December 12) heavy rain fell, which would have produced an instantaneous effect on the expectant vegetation, and perhaps most of all on the present group, throughout Sinai. On these collections Mr. Mitten writes as follows :

' Very little is yet known of the mosses of Palestine. Twelve species were enumerated by Lorentz : "Ueber die Moose die Hr. Ehrenberg in den Jahren 1820-1826 in Ægypten, der Sinai-Halbinsel und Syrien gesammelt. Aus den abhandlungen der Konigl. Akad. der Wissenschaften, Berlin, 1867." Most of these were in a barren state, and belonged to the same species as those collected by Mr. Hart, who gathered twenty-seven and four Hepaticæ. The two collections together only raise the number of species to thirty-two, of which several are, from the state of the specimens, by no means certainly determined.

1. *Anisothecium varium*, Hedw. (*Dicranum*). Gaza, barren.

2. *Grimmia apocarpa*, Linn. Jebel Mûsa, Jebel Katharîna, a barren specimen without hair-pointed leaves.

ADDITIONS TO FLORA OF PALESTINE.

3. *Grimmia leucophæa*, Grev. With the preceding, barren. This is probably the same as *G. sinaica* (Hampe), enumerated by Lorentz, who suspected his specimens to be a state of *G. leucophæa*.

4. *G. trichophylla*, Grev. Jericho, a few slender stems with very young fruit.

5. *G. pulvinata*, Linn. Jericho, with old fruit.

6. *G. crinita*. Mount Hor, with old fruit. Edom. The barren stems of this species seem to answer to the description given by Lorentz of his " Trichostomum Aronis," p. 29, and to his figures of that moss, T. V. His specimens being barren, the piliferous leaves surrounding the fruit had not been developed. In this Grimmia the transformation is considerable and rather abrupt.

7. *Hymenostylium rupestre*, Schwæg. (*Gymnostomum*). Jebel Mûsa, Jebel Katharîna, barren.

8. *H. verticillatum* (*Encladium*, B. et S.). Same locality as preceding, and in the same barren condition. It is also included in Lorentz's list.

9. *Tortula vinealis*, Brid. Jericho, barren.

10. *T. rigidula*, Hedw. (*Trichostomum*). Gaza, barren. Seems to be this species.

11. *T. unguiculata*. Gaza, a few barren stems.

12. *T. tophacea*, Brid. (*Trichostomum*). Mount Hor, Edom. Quite barren, as it is mentioned by Lorentz.

13. *T. nitida*, Lindb. (*Trichostomum*). Same locality as last.

14. *T. revoluta*. Gaza, barren stems intermixed.

15. *T. muralis*, L. Jericho, with fruit. Beersheba, Jaffa. All are in the piliferous state, as were Lorentz's specimens.

16. *T. membranifolia*, Hook. Jericho. Edom, with fruit, but old. Lorentz had it from Bîr Hummam (between Huleh and Gennesaret ?).

17. *T. inermis*, Mont. Jebel Mûsa, with old fruit ; also Edom, in a smaller state.

18. *T. ambigua*, B. et S. Gaza, Mount Hor. The specimens being fairly complete, permit this to be determined. Lorentz enumerates

T. rigida barren, and *T. rigida* is by priority the name for this species as defined by Hudson.

19. *Eucalypta vulgaris*, Hedw. Jebel Katharina. Mount Hor, with old capsules.

20. *Entosthodon Templetoni*, Schwæg. Jebel Mûsa, Jebel Katharina, with old fruit.

21. *Funaria hygrometrica*, Linn. Jerusalem, Hospital of St. John's ruins, a fertile stem with Lunularia.

22. *Bryum argenteum*, Linn. Jericho, a few barren stems.

23. *B. atropurpureum*, Wet. et Mohr. Joppa, Beersheba, Jericho. Nearly all barren, all without capsules.

24. *B. turbinatum*, Hedw. Jebel Mûsa, Jebel Katharina. Small stems, with leaves which will not revive, all quite barren, and uncertain.

25. *Hypnum Ehrenbergii*, Lorentz. (*Brachythecium*). Jericho, a few barren fragments. Lorentz's specimens were also barren, and they come from Syria. It seems to be a species very closely related to *H. lutescens*, but with its branch-leaves not so attenuate. This species has not been gathered since Lorentz's time.

26. *H. velutinum*, L. Jebel Mûsa, Jebel Katharina. Very young fruit is on the specimens, showing the rough seta.

27. *H. ruscifolium*, Neck. Jebel Katharina, a small barren tuft with leaves mostly small and deformed. Lorentz has this in his list, also *H. tenellum*, as well as *Amblystegium filicinum* and *A. serpens*.

HEPATICÆ.

1. *Fossombronia angulosa*, Raddi. Gaza. A few very small stems with young fruit, having the spores, as usual, smooth in this species.

2. *Otiona Aitonia*, Corda. Jericho. Some fragments, certainly of this genus, may be this species. An African and Indian genus. This species is found in Madeira and Socotra.

3. *Lunularia vulgaris*. Hospital of St. John's, Jerusalem, with gemmæ as usual.

4. *Riccia lamellosa*, Raddi. Beersheba. A few small fronds with the white scales more prominent than in Raddi's original specimens.'

In 1882 Mons. C. W. Barbey published a list of mosses in his 'Herborization au Levant,' which includes fifteen species from Palestine, chiefly from Beirût, Hebron, and the neighbourhood of Jerusalem. Four of these only are in my list. They are: *Dicranella varia*, *Funaria hygrometrica*, *Rhynchostegium* (*Hypnum*) *rusciforme*, and *Trichostomum nitidum*. I will take the liberty of enumerating the others, as Mons. Barbey's valuable work may not be always accessible in this country:

'*Barbula cuneifolia*, Diks. Beirût.

B. lævipila, Bridel. Solomon's Pools, near Bethlehem.

B. muralis, L. Hebron; and var. γ. *æstiva*. Jerusalem, etc.

B. unguiculata, Hedw., forma. Beirût.

B. subulata, Bridel., var. *subinermis*. Hebron.

B. vinealis, Bridel. Hebron and Solomon's Pools.

Funaria calcarea, Sch. Beirût.

Gymnostomum calcareum. N. et H., var. δ. *brevifolium*. Nahr el Kelb, Beirût.

Homalothecium sericeum, Sch. Solomon's Pools.

Rhyncostegium tenellum, Sch. Beirût.

Trichostomum Barbula, Sch. Beirût.'

Mons. Barbey also found two Hepaticæ near Beirût, *Anthoceras lævis*, Willd., and *Asterella hemisphærica*, P. Beauv.

Decaisne, in his 'Florula Sinaica,' 1834, records four mosses from Mount Sinai, *Hypnum aduncum*, Linn., *H. rusciforme*, Neck., *Bryum turbinatum*, Sw., and *Trichostomum aviculare*, Beauv. Of these the first and the last do not appear in my list. This seems to complete our knowledge of Palestine and Sinaitic mosses, and gives a total of forty-three mosses and six Hepaticæ. Of these, twelve mosses are from the Sinaitic mountains, and probably three more of Lorentz, mentioned above, without locality, are from here also; but in Mr. Mitten's opinion these new mosses of Lorentz 'appear to have been barren, and if in better state might not have been considered new.' Almost all these species (*Hypnum Ehrenbergii* and *Otiona Aitonia* being excepted) are such as are everywhere common, especially in the Mediterranean countries.

LICHENES.

The following have been determined by the Rev. James Crombie, F.L.S., having been forwarded to him by Mr. Oliver, of Kew:

Omphalaria sp. —— (?) Sterile and uncertain. On rocks, Jebel Katharina.

Ramalina crispatula, Nyl. Sterile. Bîr es Sebâ, on the ground. Distr. Canaries, North Africa.

Cladonia pyxidata, L. Sterile and but little evolute. On the ground, Jericho. Distr. cosmopolitan.

Lecanora (Squamaria) crassa, D.C. Sparingly fertile. Amongst mosses on rocks. Wâdy Ghurundel, Edom. Distr. Europe, Africa, Australasia.

Lecanora (Sarcogyne) pruinosa, Sm. On rocks of the Tîh escarpment, near Wâdy Zelegah. Distr. Europe, N. America.

Endocarpon hepaticum, Ach. On the ground. Ghôr es Safieh. Distr. Europe, Africa, N. America.

INSECTA, ETC.

to face p. 175.   PALESTINE EXPLORATION FUND.

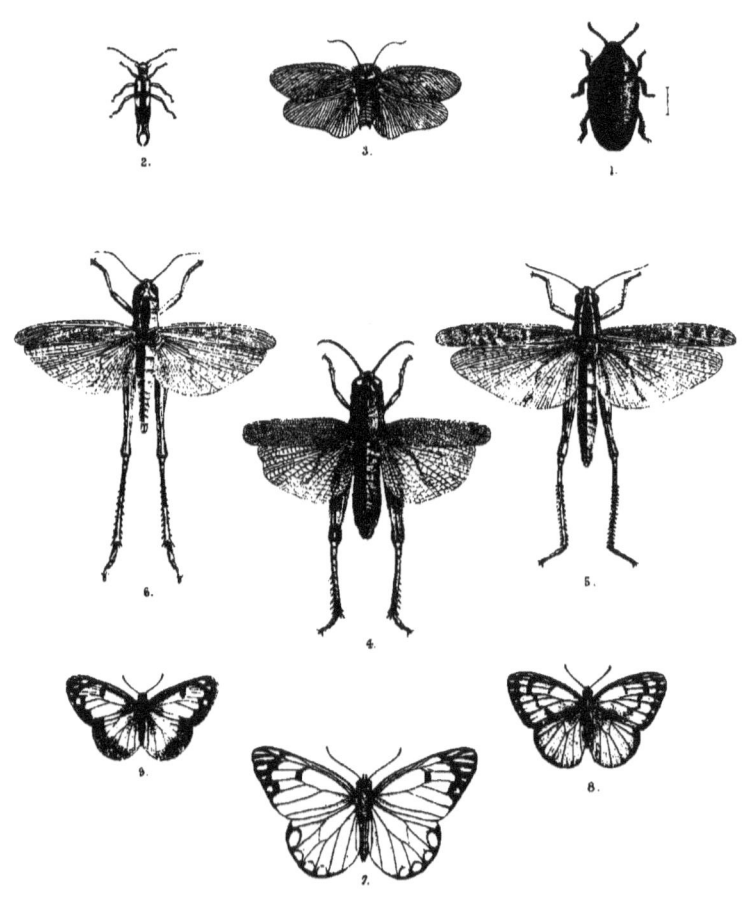

R. Mintern del et lith.
1. Gabella Harti, *Jans.*   4. Caloptenus pilipes, *Jans.*   7. Belenois Lordaca, *Walk.*
2. Forficula Lucasi, *Dohrn*   5. Cyrtacanthacris ornatipes, *Walk.* 8. Teracolus chrysonome, *Klug.*
3. Heterogamia maris-mortui, *Jans* 6. Cyrtacanthacris notata, *Walk.* 9. Teracolus phisadia, *Godt.*

INSECTA, ETC.

In the classes Arachnida, Myriopoda and Insecta, some small gatherings were made from time to time. The season was so unsuitable for these forms of life that there was very little material to work upon, and I feel that an apology may be necessary for introducing such brief and fragmentary results. My excuse must be the desire to contribute, in however small a degree, to the knowledge of the natural history of the district in all its branches, and it will be found, I believe, that my entomological captures are of interest. My specimens have been examined and determined with the utmost care by Mr. Oliver Janson, and I consider myself fortunate in having secured the services of so painstaking a worker. For the recommendation to him I am indebted to Mr. Waterhouse, of the British Museum, who also kindly determined for me some of the species mentioned, and for whose valuable advice I am very grateful.

In the preface to Canon Tristram's 'Survey of Western Palestine' will be found a general statement of the condition of our knowledge of the Arachnid and Insect Fauna of Palestine. Such knowledge as is available is merely fragmentary, nor does that author find enough material to make any important deductions therefrom. He mentions that there are five Ethiopian (Nubian) butterflies known from the Dead Sea. He does not, however, give their names.

Amongst the butterflies of the family Pieridæ, Wallace classes two genera as being confined to the Ethiopian region. Of one of these, Teracolus, I obtained two species in the Ghor. *T. phisadia*, Godt., was abundant. In the same family, *Pieris lordacea*, Walk., known only from

INSECTA, ETC.

Harkeko (lat. 15°), in Upper Nubia, was also obtained in the Ghôr. In fact, my small collection from the Ghôr exhibits in some cases very decided Ethiopian affinities, and will, at least, suffice to show the desirability of thorough investigation into the insect fauna of the Dead Sea basin.

For further information I may refer my readers to Reiche and De Saulcy's papers on the Coleoptera collected by De Saulcy in the East, Syria, etc.[1] On the beetles of Sinai, the appendix, by G. R. Crotch, on the Coleoptera collected by the Ordnance Survey, at p. 263 in the 'Report,' is the most important paper. Mr. F. Walker's papers on the insects collected by J. K. Lord, 'in Egypt, Arabia, and near the African shore of the Red Sea,' deal largely with Sinai. The Coleoptera and Hymenoptera were published separately.[2] The Diptera appeared in the *Entomologist* for March, April, and July, 1871; the Lepidoptera in the same journal for April, August, and September, 1870.

In the following list, except where otherwise stated, the species have been determined by Mr. Oliver Janson.

ARACHNIDA.

SCORPIONIDÆ.

Androctonus (*Linrus*) *quinque-striatus*, Hemp. et Ehr. Jericho; Sinai; Mount Hor. Seems to be the commonest species of scorpion. Found in Egypt and Nubia to Dongola.

SOLPUGIDÆ.

Galeodes araneoides (?), Linn. Salted remains of this species (or one extremely near it) were picked up on the margin of the Dead Sea. Mr. Waterhouse determined it to be a species of Solpugidæ, and it agrees with figures of the present well-known and savage spider. Found on the Russian and Asiatic steppes, and believed to inhabit Egypt and Arabia.

---

[1] 'Ann. de la Soc. Ent. de France,' 1855, p. 561; 1856, p. 353; 1857, p. 169, and 1858, p. 5.

[2] 'A List of Hymenoptera collected by J. K. Lord, Esq.' By Francis Walker, F.L.S. London: W. Janson, 1871. 'A List of Coleoptera,' etc., etc.

ARANEIDÆ.

Prosthesina, sp. Sinai.

Sparacis, sp. Very abundant on the portion south-west of 'Akabah, on Jebel Herteh, and elsewhere on Sinai. These spiders, grotesque, large-headed, sandy-coloured animals, were very nimble, and appeared to be in great demand amongst the members of the chat family.

Several other spiders were brought from the Ghôr, but were not in a fit condition for examination.

ACARIDEA.

Hyalomma ægyptium, Linn. Ghôr es Safieh. 'Said to be found on the Land Tortoise. Known from Egypt and Barbary.'

Argas persicus, Fisch. (?) 'The specimen is discoloured, and without limbs, and cannot be identified with certainty.' I found it parasitic on a blackbird which I shot in the Ghôr es Safieh. Found in Persia and Egypt.

MYRIAPODA.

CHILOPODA.

Scolopendra canidens, Newport. (?) 'The specimen is very young, and cannot be identified with certainty.' Mount Hor. Known from Egypt.

CHILOGNATHA.

Spirostreptus bottæ, Gerv. This repulsive-looking black milliped appears to be common along the wâdies leading to the 'Arabah from Edom. A specimen at Mount Hor was about five inches long and of the diameter of a lead pencil. Known from Tripoli, Egypt, Syria, and Abyssinia.

INSECTA.

COLEOPTERA.

SCARABÆIDÆ.

Ateuchus sacer, Linn. Found amongst the salted débris along the Dead Sea shore. Named by Mr. Waterhouse.

Bubas bubalus, Oliv. Abundant about Gaza, where I used to watch it conveying the droppings of animals below the surface, and substituting for them the soil excavated from their long vertical burrows. Common throughout the Mediterranean region.

Pachdema sinaitica, Crotch. ' *P. pilosa*, Walk., is probably identical with this species, its greater size (seven lines) being the only point mentioned in the description in which it differs. Although both names were published in the same year (1871), I have no hesitation in adopting the name given to it by Crotch in preference to that of Walker, the descriptions of the latter author being almost useless for the purpose of identification, besides many of the species being referred to wrong genera, and in some instances to wrong families.' Sinai only.

Oryctes nasicornis, Linn. Amongst the Dead Sea drift, Ghôr es Safieh. Named by Mr. Waterhouse. A Mediterranean beetle.

BUPRESTIDÆ.

Steraspis squamosa, Klug. Recorded by Walker from Mount Sinai, Cairo, and Hor, Tamanib on the Red Sea shore of Africa. Found by me in Wâdy el 'Ain on acacia-trees. Named by Mr. Waterhouse. Found in Syria and Egypt.

Galbella Harti, O. Janson. *n. sp.* This was the only living thing I found in the intensely salt marls about 600 feet above the Dead Sea in Ghôr. Where the marls were cut through by occasional small *seils* there was other life, but this little beetle inhabited the most arid and driest portions of the white chalky soil. Its nearest relative (G. beccari) is

Abyssinian, and its occurrence here points to the Ethiopian element in the life of the Ghôr, as instanced in Teracolus amongst butterflies, and perhaps Cyrtacanthacris amongst locusts.

TENEBRIONIDÆ.

Adesmia carinata, Sol. Sinai. Known from Persia. Not in Walker's or Crotch's Sinai list.

Adesmia lacunosa, Klug. Jebel Wattyeh. Known from Cairo, and other parts of Sinai. Named by Mr. Waterhouse.

Adesmia tenebrosa, Sol. With the last. There are about ten species of this genus found in Sinai. Named by Mr. Waterhouse. The present species is not in Walker's or Crotch's list. It is found in Persia.

Acis elevata, Sol. 'Ayûn Mûsa. Found in Egypt. Named by Mr. Waterhouse.

Prionotheca coronata, Oliv. Wâdy Ghurundel ('Elim'). Named by Mr. Waterhouse. Found as far south as Nubia in lat. $15°$.

Ocnera hispida, Fask. Wâdy el 'Ain. Found also in Egypt. Named by Mr. Waterhouse.

Mesotena punctipennis, Sol. Sinai. Not in Walker or Crotch. Known from Egypt and Algeria. An allied species, M. oblonga, is recorded by Crotch from Sinai.

Scleron orientale, Fab. Wâdy el 'Ain. Not in Walker or Crotch. It is known from Egypt, Syria, and India.

Himatismus villosus, Haag. Sinai. Known from Egypt. Not in Walker's or Crotch's Sinai list of beetles.

COCCINELLIDÆ.

Epilachna chrysomelina, Fab. Wâdy Ghurundel, Edom, and Ghôr es Safieh. Known from Europe, Arabia, Persia, West and South Africa.

HYMENOPTERA.

APIDÆ.

Anthrophora nidulans, Fab. I dug this bee out of a burrow in a sandy bank in the Wâdy 'Arabah between Sinai and Palestine while gathering bulbs. It is not included in Walker's Sinai list. It inhabits Southern Europe, Algeria, and Madeira. My visit was too early for the vast majority of these insects.

VESPIDÆ.

Vespa orientalis, Linn. This handsome hornet was the only conspicuous insect of the family that was frequently to be seen during the winter months at Sinai. In some places, as about the well at Wâdy Nasb, it was very abundant. Inhabits Southern Europe, Egypt, Arabia, India.

The test of a species of *Polistes* apparently was found amongst the Dead Sea drift.

DORYLIDÆ.

Dorylus juvenculus, Schnk. This swift, long-bodied wasp was several times seen in Sinai. I captured one at the southern end of Wâdy el 'Ain. Inhabits Barbary and Tripoli. Not included in Walker's list.

FORMICIDÆ.

Camponotus compressa, Fab. This appears to be a very common species of black ant in Sinai. Specimens from several ant heaps have been referred to it by Mr. Janson. It is not included in Walker's list. Inhabits Syria, India, and China.

Aphænogaster structor, Latr. Wâdy 'Arabah and 'Ayûn Mûsa. Inhabits Europe, Syria, and Palestine. Found at Cairo also by Mr. Lord. (See Walker's list.)

Polyrhachis seminiger, May. These ants, in small parties, inhabited little nests or pockets of vegetable matter agglutinated together, and appended to the branches of the tamarisks close to the shore of the Dead

Sea. This genus is tropical, chiefly Indian, and no species was captured by Mr. Lord according to Walker, so that this outlying station is very interesting. Mr. Janson informs me that there was an unnamed specimen of this ant in the British Museum.

LEPIDOPTERA.

NYMPHALIDÆ.

Pyrameis cardui, Linn. Wâdy es Sheikh in Sinai, in a clump of tamarisks. 'This well-known butterfly is found in nearly all parts of the world except Australia, where it is replaced by the closely-allied *P. Kershawi*, McCoy.' Mr. Lord captured it at Cairo and in Upper Egypt.

PIERIDÆ.

Pieris (*Belenois*) *lordacea*, Walk. Ghôr es Safieh at the south end of the Dead Sea. Described from specimens collected by Mr. Lord at Harkeko, on the African shore of the Red Sea, in lat. 15°. Mr. Walker states that this butterfly belongs to the group of which *P. mesentina*, Cram. (also an Ethiopian form) is the type, and has much affinity with that species. The two were in company at Harkeko, and it will be interesting to know if P. mesentina also occurs in the Ghôr. A characteristic example of the tropical conditions of the Ghôr.

Teracolus phisadia, Godt. These butterflies were just emerging from the pupa state, which were attached in great abundance to the twigs of a species of Rhamnus by the edge of the river Hessi in the Ghôr. Known only from Aden and Abyssinia twenty degrees further south.

T. chrysonome, Klug. Ghôr es Safieh. Known from Ambukohe in Somali, fifteen degrees south of the Ghôr. Wallace, in his ' Distribution of Animals ' (ii., 478), states that two genera of Pieridæ, of which the present is one, are confined to the Ethiopian region. Teracolus is known by about ten species from Madagascar, Arabia, and S. Africa. These last three butterflies offer satisfactory illustration of the view of the past northern range of tropical forms along the Red Sea and the 'Arabah to the Ghôr, now interrupted by the aridity of the climate. None of these were captured by Mr. Lord or myself in Sinai.

INSECTA, ETC.

CRAMBIDÆ.

Stenopteryx hybridalis, Hub. Sinai. Ranges over the whole warm world almost, except Australia.

N.B.—Species of Lasiocampidæ, Noctuidæ and Agrotis were also captured in Sinai, but they were not in good condition. A very regularly constructed caterpillar-case, made of small twigs and not unlike those of the larvæ of species of caddis-worm, was gathered in Wâdy Zelegah in Sinai. It belonged to some species of Psyche.

DIPTERA.

HIPPOBOSCIDÆ.

Hippobosca camelina, Savign. This camel tick was frequently found in Sinai. It is known from Egypt.

NEUROPTERA.

TERMITIDÆ.

Holotermes vagans, Hag. These white ants (chiefly the males (?)) appeared in some abundance in the 'Arabah in December, where they were greedily devoured by the 'hopping thrush,' *Argya squamiceps*, Rupp. The insect is known from Persia, and, with many other forms of life, found on either side, east and west, of Arabia; no doubt inhabits also that little-known desert region.

ORTHOPTERA.

FORFICULIDÆ.

Forficula lucasi, Dohrn. In the Edomitic wâdies leading to the 'Arabah. Inhabits Syria and Egypt.

BLATTIDÆ.

Heterogamia maris-mortui, O. Janson. *n. sp.* Captured in the Ghôr es Safieh, being attracted to the tent lights at night. A light-coloured yellowish-brown cockroach, with prettily reticulated wings.

ACRIDIIDÆ.

Cyrtacanthacris ornatipes, Walk. In the Ghôr es Safieh, at the south end of the Dead Sea. 'Described from a single specimen in the collection of the British Museum, the locality of which was unknown.'

Cyrtacanthacris notata, Walk. Ghôr es Safieh. 'One specimen, which appears to be merely a variety of this species, differs in its much lighter colour and small size, the expanse of wings being only 16 lines.' Inhabits Upper Egypt. This considerable genus of grasshoppers appears to be mainly known from the tropical parts of America and Africa.

Acridium tataricum, Roes. Jericho. 'Known from Asia Minor, Europe, Africa.'

Caloptenus pilipes, O. Janson. *n. sp.* Ghôr es Safieh.

Œdipoda cærulans, Roes. Sinai. 'Known from S. Europe, Egypt, Asia Minor, Madeira, S. Africa.'

Œdipoda insubrica, Savigny. Gaza. This locust inhabits S. Europe, Syria, and Algeria.

Pœcilocera bufonia, Klug. Wâdy Berrâh in Sinai; also very conspicuous on the 'Osher' (*Calotropis procera*) in the Ghôr es Safieh. A blackish purple, strong, uncanny-looking, locust-like insect, two or three inches long, with small yellow spots. It is known from Egypt and Syria.

Tryxalis unguiculata, Ramb. I captured this insect first on Jebel Watlyeh in Sinai, and afterwards in the Ghôr es Safieh. It closely resembled the dried twigs and straws which it lived amongst. 'Known from S. Europe, Syria, Nubia and N. India.'

RHYNCHOTA.

LYGÆIDÆ.

Lygæus militaris, Stal. Frequent about Mount Hor and other parts of Edom, in Arabia Petræa. A brightly-coloured red and black little land bug.

PYRRHOCORIDÆ.

Pyrrhocoris apterus, Geoff. Same locality as last. This little 'red bug' is found from England across Europe to Syria.

The following descriptions are by Mr. Oliver Janson, with whom the type specimens are deposited:

Galbella Harti, O. Janson. Elongate-ovate, moderately convex, blue-black, shining. Head convex, with a large shallow fovea in front, coarsely and rather closely punctured, a small space just behind the fovea smooth; antennæ strongly serrate from the fifth joint, joints 7-11 twice as broad as long. Thorax coarsely punctured, the punctures closer and somewhat confluent at the sides, slightly impressed near the anterior margin, the basal margin bisinuate, posterior angles acute. Scutellum triangular, smooth. Elytra four times as long as the thorax, strongly punctate-striate, with alternate rows of simple punctures near the suture, the outer disk with coarser confluent punctures, the interstices of which form transverse ridges; base, sides, and apex confusedly punctured; humeral callus strong and smooth. Underside strongly punctured; metasternum with a slight central carina; sides of the abdomen finely serrate. Legs short, tibiæ strongly dilated, the outer edge rounded and finely serrate. Length, 6 mm., width 2½ mm.

Closely allied to *G. beccari*, Gest., but differs chiefly from that species in the sculpture of its elytra and in having a distinct fovea on its head.

The present species is the fifth known of this interesting and peculiar genus (originally referred by Westwood to the *Eucnemidæ*), the others being *violacea*, Westw., from India; *beccari*, Gestr., from Abyssinia; *felix* (*Zanthe*), Mars., from Cyprus; and *cærulea* (*oncomæa*), Saund., from Siam.

Heterogamia maris-mortui, O. Janson. ♂. Pale testaceous, shining. Head pale yellow in front, the base and eyes black; antennæ nearly as long as the body, ferrugineous. Thorax transverse-elliptical, sparsely punctured, slightly strigose at the base, the entire surface clothed with sparse short rufous hairs, the anterior margin fringed with longer hairs, the disk impressed on each side and marked in the centre with a somewhat anchor-shaped ferrugineous spot, the sides semi-pellucid and whitish. Mesonotum uneven, impunctate, ferrugineous brown, with the margins pale; scutellum black. Metanotum elevated in the centre, uneven, impunctate, sparsely pubescent, and with obscure brown spots. Elytra pellucid, with numerous irregularly-disposed brown spots; costal area whitish, the costa fringed with short rufous hairs. Wings pellucid, the veins brown, abdominal margin white. Abdomen above finely strigose, with a slight elevated median line, apex ferrugineous. Legs with sparse rufous hairs; tibial spines long, rufous-brown.

Length of the body, 14 mm.; length of elytra, 19 mm.; length of thorax, 4½ mm.; width of thorax, 7 mm.

Appears to be most nearly allied to *H. conspersa*, Brunn. (*syriaca*, Sauss.).

Caloptenus pilipes, O. Janson. Robust, tawny brown. Head convex, and mottled with lighter brown behind; vertex sub-elliptical, concave; face sparsely punctured, shining; mouth, margins of the frontal costa and sides of the face whitish; eyes luteous; antennæ testaceous. Thorax produced and broadly rounded behind; the base and sides finely scabrose; the central carina well-marked and entire, the outer ones broken and almost obsolete behind; with four transverse impressed lines, the first close to the anterior margin, well-marked at the sides, but almost obsolete on the disk, the second sinuous and only reaching a little beyond the outer carinæ; the third sinuous and extending nearly to the lateral margins; the fourth straight and of the same length as the third. Elytra pale cinereous brown, the base, veins, and several obscure marks near the apex darker. Wings pellucid and slightly tinged with purple, which is more pronounced in certain lights; veins dark brown. Anterior and intermediate legs testaceous, sparsely pubescent, underside and tarsi whitish; posterior femora very broad, clothed with fine white pubescence, outer side pearly white, inner side red; posterior tibiæ pubescent, bright yellow, paler towards the base; apex of spines black. Abdomen and underside pale ferrugineous brown, shining, apex of the former pale cinereous; prosternal spine strong, rounded, obtuse at its apex.

Length of the body, 35 mm.; length of elytra, 24 mm.; length of the posterior femora, 20 mm.; length of the posterior tibiæ, 19 mm.

Allied to *C. testaceus*, Walk.

MOLLUSCA.

MOLLUSCA.

MOLLUSCA FROM 'AKABAH, RED SEA.

The numbers after the species refer to Issel's 'Malacologia del Mar Rosso.' Where there is no number, the name McAndrew, Beke, or Burton refer to the lists in these gentlemen's works respectively, the earliest alone being mentioned.

Where the species appears to be now first recorded, I have endeavoured to give its range from Reeve's 'Monographs,' Kiener, or Weinkauff.

PTEROPODA.

* *Hyalæa uncinnata*, Raug. Atlantic.
  ,, *quadritentata*, Lessur. (McAndrew).

GASTEROPODA.

Murex anguliferus, Lamk. (261). Also at 'Ayûn Mûsa, Gulf of Suez.

Fusus polygonoides, Lamk. (303). And at 'Ayûn Mûsa.

Pleurotoma cingulifera, Lamk. (335).

Triton lampas, Linn. (283, ' raro ').
  ,, *trilineatus*, Reeve. (McAndrew).
  ,, *pilearis*, Linn. (Milne in Beke).
  ,, *rubecula*, Linn. (McAndrew).

Triton anus, Linn. (284).
  ,, *aquatilis*, Reeve. (McAndrew).
* *Ranella granifera*, Lamk. Philippines.
  ,, (*Lampas*) *affinis*, Brod. (McAndrew).
* ,, *rhodostoma*, Bech. Philippines.
Phos roseatus, Hinds. (McAndrew).
Nassa gemmulata, Lamk. (247).
* ,, *sordida*, A. Adams. Philippines.
* ,, *delicata*, A. Adams. Philippines.
* ,, *plebecula*, Conrad.
Purpura serta, Lamk. (236 ' rara ').
Ricinula (*Pentadactylus*) *ablolabris*, Blainv. (McAndrew).
  ,,   ,, *fiscillum*, Chemn. (McAndrew).
  ,,   ,, *spectrum*, Reeve. (McAndrew).
  ,, (*Purpura*) *tuberculata*, Blainv. (241).
Coralliophila (*Rhizochilus*) *costularis*, Blainv. (McAndrew).
  ,,   ,, *exarata*, Pease. (McAndrew).
  ,,   ,, *madreporianus*, A. Adams. (McAndrew).
Latirus (*Turbinella*) *turritus*, Linn. (McAndrew).
* *Mitra variegata*, Reeve.
* ,, *cucumerina*, Lamk. Pacific Islands.
* ,, *digitalis*, Chemnitz. Philippines.
* ,, *filosa*, Carpenter. Philippines.
* ,, *consanguinea*, Reeve.
  ,, *osiridis*, Linn. (McAndrew).
* ,, *abbatis*, Chemnitz. Philippines.
* ,, *modesta*, Reeve. Philippines.
Columbella (*Ricinula*) *mendicaria*, Lamk. (242).
* ,, *ligula*, Duclos.
  ,, *conspersa*, Gask. (McAndrew).

* *Columbella albinodulosa*, Gask. Philippines.
  ,, *albina*, Kiener. (McAndrew).
* *Harpa minor*, Rumph. Philippines.
* *Cassis saburon*, Lamk. Atlantic and Mediterranean.
  ,, *torquata*, Reeve. (231).
Natica mamilla, Linn. (457).
  ,, *melanostoma*, Gmel. (458).
  ,, *marochiensis*, Gmel. (McAndrew). Canaries.
Terebra crenulata, Linn. (260).
  ,, *babylonica*, Lamk. (McAndrew).
  ,, *affinis*, Gray. (McAndrew).
* *Solarium cingulum*, Kiener. Sandwich Islands; Philippines; Japan.
Conus arenatus, Brug. (322). And at 'Ayûn Mûsa.
  ,, *nussatella*, Linn. (McAndrew).
* ,, *nigropunctatus*, Sowerby.
  ,, *tesselatus*, Brug. (324).
* ,, *generalis*, Linn. Ceylon.
* ,, *nanus*, Brod. Lord Hood's Island; Pacific.
  ,, *vicarius*, Lamk. (McAndrew).
* ,, *mustellinus* (*sulphuratus*, Brug.). Philippines.
* *Strombus auris-dianæ*, Linn. Philippines.
  ,, *gibberulus*, Linn. (274). And at 'Ayûn Mûsa.
  ,, *floridus*, Lamk. (270).
  ,, *fusiformis*, Sowerby. (273).
* ,, *plicatus*, Lamk. (*S. dentatus*, Linn.). Philippines.
Cypræa turdus, Lamk. (196).
  ,, *isabella*, Linn. (201).
  ,, *erosa*, Linn. (204).
* ,, *tabescens*, Solander. Tahiti; Ceylon; Pacific.

MOLLUSCA.

* *Cypræa gangrenosa*, Dillwyn. China.
,, *globulus*, Linn. (Beke). East Indies.
,, (*Trivia*) *oryza*, Lamk. (Pagenstecker in R. Kossman).
* *Cerithium fenestratum*, Sowerby.
,, *lacteum*, Philip. (McAndrew).
,, *Kochi*, Philip. (McAndrew).
,, *tuberosum*, Fabr. (340). And at 'Ayûn Mûsa.
,, *moniliferum*, Kiener. (352). ,,
,, *echinatum*, Lamk. (McAndrew).
,, *rostratum*, Sowerby. (McAndrew).
Modulus tectum, Gmel. (Milne in Beke).
Turritella trisulcata, Lamk. (505).
Nerita polita, Linn. (517).
,, *rumphii*, Reclus. (Pagenstecker in Kossman).
Turbo margaritaceus, Linn. (Milne in Beke).
,, *sp.* (?) too young, and *sp.* (?) *opercula*.
Trochus erythræus, Brocchi. (534). And at 'Ayûn Mûsa.
,, *dentatus*, Forsk. (528). ,,
Clanculus Pharaonis, Linn. (542).
Forskahlea declivis, Forsk. (541). And at 'Ayûn Mûsa.
* *Omphalius cariniferus*, Bech.
Monodonta obscura, Wood. (McAndrew).
* ,, *australis*, Lamk.
Perrinea (*Euchelus*) *stellata*, A. Adams. (McAndrew).
Gena lutea, Linn. (McAndrew).
Cerithiopsis clathrata, A. Adams. (McAndrew).
Phorus (*Triphorus*) *corrugatus*, Reeve. (McAndrew).
Phasianella jaspidea, Reeve. (520, ' non commune '). And at 'Ayûn Mûsa.
* *Subemarginula imbricata*, A. Adams.

Haliotis scutulum, Reeve. (McAndrew).
Hipponyx australis, Quoy. et Gaim.
* *Patella variabilis*, Krauss. Valparaiso, Reeve. This species was determined by Mr. Edgar Smith of the British Museum.
Planaxis Savignyi, Desh. (481).
* *Siphonaria cochleariformis*, Reeve. China.
*   ,,   *lecanium*, Philip.
Bulla ampulla, Linn. (413). And at 'Ayûn Mûsa.
Atys (alicula) cylindrica, Chemnitz. (418).
,,   ,,   *succisæ*, Ehr. (420). And at 'Ayûn Mûsa.
* *Buccinulus solidulus*, Linn.
* *Janthina nitens*, Manke. Mediterranean.

CONCHIFERA.

Tridacna elongata, Lamk. (93).
*   ,,   *crocea*, Lamk.
* *Lima hians*, Gmelin. Mediterranean and Atlantic.
Pinna, sp. (?) Too much damaged.
Pecten sanguinolentus, Gmel. (156, 'raro').
,,   *lividus*, Lamk. (165).
Chama cornucopia, Reeve. (89). And at 'Ayûn Mûsa, Gulf of Suez.
* *C. obliquata*, Reeve. Philippines. And at 'Ayûn Mûsa.
Spondylus, sp. Not named by Smith or Sowerby. Probably new, but specimens worn.
Ostræa frons, Linn. (McAndrew).
* *Arca (Barbatia) divaricata*, Sowerby. Pacific Islands.
*   ,,   ,,   *barbata*, Linn. Mediterranean and Atlantic.
*   ,,   (*anomalocardia*) *scapha* (?), Chemnitz (113).
* *Pectunculus lineatus* (?), Reeve. West Indies.
,,   *Siculus*, Reeve. Sicily; Mediterranean.

Pectunculus pectiniformis, Lamk. (110).   And at 'Ayûn Mûsa.
* *Cardita antiquata*, Reeve.   Ceylon.
  „   *variegata*, Brug. (96, ' non commune ').
Modiola (*Perna*) *auriculata*, Krauss. (McAndrew).
* *Venus toreuma*, Gould.   Mangse Island.
  „   (*Chione*) *costellifera*, Ad. et Reeve. (McAndrew).
  „   *reticulata*, Linn. (47).
Circe arabica, Chemnitz. (53).
  „   *lentiginosa*, Chemnitz. (54).
* „   *tumefacta*, Sowerby.
* *Artemis rubecundus*, Philip.
* *Cardium fragum*, Linn.
Tellina scobinata, Linn. (34).
Paphia (*Mesodesma*) *glabrata*, Deshayes. (18).
Lucina dentifera, Jonas. (101).
Cytherœa florida, Lamk. (63).

FROM AIN MUSA, GULF OF SUEZ.

GASTEROPODA.

Murex tribulus, Linn. (Smith in Burton).   Eastern Seas.
Strombus tricornis, Lamk. (267).
  „   *lineatus*, Lamk. = *S. fasciatus*, *Born.* (McAndrew).
Ancillaria striolata, Sowerby. (McAndrew).
Pteroceras truncatus, Lamk. (Smith in Burton ; Kiener).
Conus quercinus, Linn. (330).
Cerithium cæruleum, Sowerby (346).
  „   (*Pirenella*) *Caillandi*, Potiez. (355).   Determined by E. Smith.
* *Nerita antiquata*, Reclus.   Philippines.

FROM AIN MUSA, GULF OF SUEZ.

Nerita albicilla, Linn. (511).
* ,, *stella*, Chemnitz. Determined by E. Smith.
  ,, *marmorata*, Reeve. (McAndrew). Determined by E. Smith.
Pyrula paradisiaca, Martini. (317).
Fissurella Ruppellii, Reeve. (551). Determined by E. Smith.
Marginella, sp. (?). Not named by Smith or Sowerby. Probably new.

CONCHIFERA.

Vulsella spongiarum, Lamk. (146).
* *Chama iostoma*, Conrad. Sandwich Islands.
* ,, *corrugata*, Brod. Central America, Reeve.
Arca retusa, Lamk. (117).
  ,, (*Barbatia*) *fusca*, Brug. (118).
* *Circe abbreviata*, Lamk.
  ,, *crocea*, Gray. (72).
  ,, *pectinata*, Linn. (67).
Pectunculus arabica, H. Adams. (McAndrew.)
Mytilus variabilis, Issel. (128).
Plicatula ramosa, Lamk. (168).
* *Avicula* (*Meleagrina*) *ala-perdicis*, Reeve.
Mactra olorina, Philip. (13).
* *M. semisulcata*, Desh.
Lucina semperiana, Krauss. (102).
* *Diplodonta rotundata*, Turton. Atlantic and Mediterranean. Determined by E. Smith.

FROM THE MEDITERRANEAN AT JAFFA.

Columbella rustica, Linn. In enormous quantities, from which it is difficult to find other sorts on the beach at Jaffa.

Mitra corniculata, Linn.

,,   *ebenus*, Lamk.

Conus mediterraneus, Brug. Common.

Pollia bicolor, Cantr. Not in Weinkauff's list of European sea-shells, unless a synonym.

Pollia Orbignyi, Payrandeau.

Nassa incrassata, Müller.

,,   *Cuvieri*, Payrandeau ($=N.$ *variabilis*, Philip.).

Cerithium mediterraneum, Desh.

,,   *scabrum*, Olivier.

,,   *levantinum, n. sp.* Mr. Edgar Smith has kindly favoured me with the following description :

Testa pyramidalis, subturrita, alba, nigro-fusco transversim punctata. Anfract. 8 (?) superne concavi, ad latera convexiusculi, costis, longitudinalibus lirisque spiralibus cancellati, striisque inter liras sculpti. Costæ circa 11, mediocriter crassæ, paucis hic illic cæteris majoribus. Liræ spirales nigro punctatæ, circa 4 in anfr.-penult., in ult. ad 8. Anfractus ultimus ad latum sinistrum varice albo marginatus, prope aperturam leviter ascendans. Apertura parva, ovata, inferne brevissime et oblique caniculata, intus alba, nigro punctata. Labrum varicosum, intus liratum ; columella in medio leviter arcuata, superne lira intrante munita.

This species is pyramidal, somewhat turreted, moderately solid, white, and ornamented with series of transverse blackish dots upon the spiral liræ. The whorls are probably about eight in number, narrowly excavated below the suture, then a little convex and sculptured with strongish longitudinal ribs and spiral liræ, between which are still finer striæ. The ribs are ten or eleven in number on the penultimate whorl, one or two of which (varices) are white and larger than the rest. Of liræ there are four on the upper volutions, the two central ones being most conspicuous, and at the points of contact with the costæ produce a nodulous appearance. The two others, which are much less distinct, and in some specimens almost obsolete, are situated above and below at the suture. The body-whorl has about eight liræ dotted with dark brown, and a conspicuous white oblique varix on the left side opposite the labrum, near which it ascends slightly. The aperture is small, ovate, white, exhibiting the external dotting some distance from the margin, and terminates in a very short oblique canal beneath. The columella is incurved at the middle, covered with a thin callosity, and bears at the upper part a fine transverse liræ running some distance within the aperture. The labrum is white, thickened exteriorly, and finely lirate within.

Length 12 millim., diam. $4\frac{1}{2}$, aperture $3\frac{2}{3}$ long, 2 wide.

This species has been kindly examined by the Marquis Monterosata, to whom it is unknown as a Mediterranean form, and there does not appear to be any species from that region sufficiently resembling it with which a comparison can be made.

Besides the black dots already mentioned, other paler specks are discernible under a lens, irregularly scattered over the surface.

Trochus (Gibbula) adriatica, Philip.

Natica Josephinia, Risso.

Cypræa pulex, Solander.

Murex cristatus, Broc.

Bulla striata, Brug.

Pectunculus glycimeris, Linn.   Forming the entire beach near Gaza.

Tapes geographica, Chemn.

Arca (Barbatia) barbata, Linn.

The above species have been, except where otherwise stated, determined by Mr. G. B. Sowerby, of Russell Street, London. I am deeply indebted to Mr. Edgar Smith, of the British Museum, for examining several critical species, and for giving me references to several of the works which I have made use of. Of these the earliest and most important relating to the Red Sea mollusca is that of Professor Arthur Issel, of Genoa, entitled, 'Malacologia del Mar Rosso,' Pisa, 1869. I have taken this work as covering and including all previous researches upon the subject. The monographs of Reeve, subsequent to 1869, have been referred to, and the whole series has been searched for information relative to the distribution of those species now first included in the Red Sea fauna. Kiener's works in the Conchylien Cabinet have been also consulted.

Next to Issel's, the most important memoir devoted to the conchology of the Red Sea is that of Mr. R. McAndrew 'On Testaceous Mollusca obtained in the Gulf of Suez,' published in the *Annals and Magazine of Natural History*, Dec., 1870.

Pagenstecker's account of the shells collected by Robe Kossman in the Red Sea has been referred to.

No doubt other scattered papers exist. The above references have been communicated to me by Mr. Smith.

In addition, in Burton's 'Land of Midian,' vol. ii., pp. 119-121, 1879, will be found a list of shells (determined by the able conchologist of the British Museum) collected on the Midian shore. In Dr. Beke's 'Sinai in Arabia,' 1878, there is an appendix containing a list of shells collected at 'Akabah, and determined by Mr. John Milne, F.G.S. These have been gone through and summarized.

My lists from 'Akabah and 'Ayûn Mûsa contain 166 determined species, and about half a dozen unnamed. Of these no less than fifty-six appear to be additional to the fauna of the Red Sea as exhibited in the above-mentioned works. Several of these will, perhaps, prove to be synonyms for other known Red Sea forms, but I have endeavoured, as far as lay in my power, to avoid falling into this error. These additions are distinguished by a prefixed asterisk.

The extraordinary profusion and variety of life in the Red Sea is well-known, and has been already commented upon. It is all the more remarkable when contrasted and brought face to face with the barrenness alongside—the poverty in all forms of terrestrial existence which its coasts exhibit.

The continuity of shallow seas and margins of continents in high temperatures and somewhat similar latitudes no doubt explains the enormous range of the marine species inhabiting the eastern hemisphere. The majority of the shells of the Red Sea are found also in the Pacific.

Many shells which are found on Samoa and Tahiti occur at 'Akabah. This at first seems strange when we consider the almost complete difference which exists between the fauna of the opposite coasts of the Atlantic, but the above consideration explains it to a certain extent.

The extreme dissimilarity between the faunas of the Red Sea and the Mediterranean has been a subject of careful consideration by Issel in the first part of his work already quoted. He finds eighteen species of molluscs common to both seas, and thirty which are equivalent to each other, or may be regarded as recent deviations from a parent stock. From a comparison of the fossilized raised beach deposits of either seas he finds that these seas were united in the Pliocene and Miocene times, and contained a common fauna. If we accept this conclusion, which is probably correct, we may view this ancient fauna as much more nearly related to that now existing in the Pacific than that in the Mediterranean; decreased temperature during glacial times and since has rendered the latter sea unfit for Pacific, warm-sea animals, which have, in the main, long become extinct there.

At the same time, the hardier Atlantic forms will have continually pressed eastwards, and exterminated the lingering pre-glacial types. A later union of post-glacial or glacial times between the two seas is believed to have taken place, but this was, no doubt, of a shallow nature over a wide sandy flat, and insufficient for any considerable interchange of life. Those few common species (to which I have several to add) may not unlikely date their advent to the Mediterranean from this most recent geological period. And no doubt the Suez Canal is even now assisting in this process. Some of my additions (Hyalæa, Ianthina) are of a nature which would enable them readily to pass from one sea to the other by this existing means.

My list contains the following species known to exist in the Mediterranean, but not before taken in the Red Sea:

Hyalæa uncinnata, Brug.
Cassis saburon, Lamk.
Ianthina nitens, Meuke.

MOLLUSCA.

Lima hians, Gm.
Arca barbata, Linn.
Pectunculus siculus, Reeve.
Diplodonta rotundata, Turton.

In addition to these the following three occur in McAndrew's paper: *Volvula acuminata*, Brug.; *Pecten varius*, Linn., and *Radula inflata*, Chemn., found by him at Suez, and stated there to occur in the Mediterranean. *Pecten pes-felis*, Linn., a Mediterranean species also, was found, according to Milne, by Dr. Beke at 'Akabah. These eleven species, added to Issel's eighteen, give twenty-nine shells common to the Mediterranean and Red Seas. It is probable also that Natica asellus and N. marochiensis, one or both, occur in the Mediterranean, since they are found at the Canary Islands, and also in the Red Sea. Neither are, however, included in my edition of Weinkauff (1873).

I have extracted from the lists in Burton and Beke's writings respectively the names of the species which do not appear either in Issel or McAndrew, and I have appended a note of their distribution as far as given in Reeve or Kiener.

GASTEROPODA.

Conus textile, Linn. (Burton and Beke). Ceylon; Society Islands.
  „ *hebræus*, Linn. (Burton). Ceylon.
  „ *ceylanensis*, Hwass. (Burton). Ceylon; Philippines.
  „ *monile*. (Beke). Ceylon.
  „ *pennacens = C. omaria*, Reeve. (Beke). Ceylon.
Terebra dimidiata, Linn. (Burton). Philippines.
  „   (*Impages*) *cærulescens*, Lamk. (Burton). Philippines.
  „   „   *nubeculata*. (Beke).
Murex inflatus, Lamk. = *M. ramosus*, Reeve. (Burton). Eastern Seas.
Nassa coronata, Lamk. (Burton). Madagascar.
Harpa solida, A. Adams. (Burton).

Triton maculosus, Reeve. (Burton). Philippines.
Solarium perspectivum, Linn. (Burton). Amboyna ; Australia.
Cypræa scurra, Chemn. (Burton). Staines Island, Pacific.
Cerithium nodulosum, Brug. (Beke). India and Moluccas.
Trochus (Polydonta) sanguinolentus, Chemn. (Burton).
Ricinula Morio. (Beke). (? *R. Morus*, Lamk.) Lord Hood's Island.

CONCHIFERA.

Tellina staurella, Lamk. (Burton). Philippines.
Cardium leucostoma, Boiss. (Burton). Singapore.
Tridacna gigas, Kinn. (Burton). Pacific and Indian Oceans.
Mactra decora. (Beke).
Pecten pes-felis, Linn. (Beke). Mediterranean.
Arca antiquata. (Beke).
Pectunculus paucipictus. (Beke).
Venus crispata. (Beke).
Cytheræa blanda. (Beke).

The fact that each collector, even though working on the beach alone, returns with additional species, shows that there is no reason to consider we have as yet arrived at the total. Mr. McAndrew remarks that up to the very last he never returned without having added several species to his collection. The following numbers represent the known species from the Red Sea:

Issel	... 640
McAndrew, not in Issel	... 355
Burton, not in Issel or McAndrew ...	... 16
Beke, not in Issel, McAndrew or Burton	... 15
Hart, not in any above	... 56
Total Red Sea molluscs	... 1082

The species in this total, common to the Mediterranean and Red Sea, form a percentage of less than three.

In addition to the above total, Mr. McAndrew collected a number of species, which are as yet undescribed, from the Gulf of Suez. These are being dealt with in the current numbers of the *Annals and Magazine of Natural History*.

LAND AND FRESH-WATER MOLLUSCA.

These have been kindly named for me, with the exception of half-a-dozen, by Canon Tristram. The other six I have had determined by Mr. G. B. Sowerby.

Sinai is poorly supplied with molluscs, and it was not till I reached Petra and crossed the 'Arabah watershed that any variety appeared.

A good gathering was made amongst drift shells on the margin of the Dead Sea, a little east of the swampy embouchure of the Tufileh river. All were dead and beached, but often in good preservation, being killed the instant they reach the salt sea. These may, I think, be all safely referred to the Ghôr es Safieh. The stream comes from the mountains and over rocks immediately into the Ghôr, where it has a quiet course of a few miles. The shells I speak of are too fragile to have travelled, except a very short distance, and are accompanied by other Ghôr deposits.

Succinea elegans, Risso. This species (determined by Mr. Sowerby) was very abundant in a dead state in the drift on the south-western edge of the Dead Sea. Canon Tristram does not mention this species as having been found in Palestine.

Helix joppensis, Roth. Ghôr es Safieh and Judæan plain west of the Ghôr. In the Wâdy 'Arabah, between 'Ayûn Buweirdeh and the Ghôr. Canon Tristram writes me, 'Probably only a variety of H. cæspitum, Drap.' It is much larger. It is in his list, but without locality.

H. syriaca, Ehr. Gaza.

H. cæspitum, Drap. Petra and Wâdy 'Arabah, near the Ghôr. Found in the north of Palestine near the coast.

H. hierochuntina, Roth. By the Jordan, below Jericho.

PALESTINE EXPLORATION FUND, PL. XII.

A
H. TUBERCULOSA, CONRA.
SIDE VIEW.   P. 203.

A
H. TUBERCULOSA, CONRA
DORSAL VIEW.   P. 203.

B
HELIX LEDERERI, PFR
P. 203.

G
ACHATINA BRONDELI, BOURG.
P. 204.

C
HELIX SPIRIPLANA. OLIV.
P. 203.

M
TELPHUSA FLUVIATILIS. SAVIGN.
PP. 56, 210.

D
CERITHIUM LEVANTINUM. SP. NOV.
HART, PALEST. EXP. FUND, 1891.
P. 196.

H
FERRUSACIA THAMNOPHILA, BOURG.
P. 204.

F
BULIMUS HARTI, SP. NOV.
HART, PAL. EXP. FUND. 1891.   P. 204.

E
CORBICULA SAULCYI, BOURG.
P. 205.

Helix protea, Ziegl. Plain of Judæa, at Bîr es Sebâ, Tell el Milh, etc.; Gaza. Very variable in colouration.

H. vestalis, Parr. Gaza.

H. tuberculosa, Conrad. First found by Professor Hull, and again in two different places by me in the plain of Judæa, between Tell el Milh and Tell Abu Hareireh. A very peculiar trochiform species, with the whorls prettily striated and mitred, giving a spirally laminated appearance. Canon Tristram speaks of it as very rare, and 'the most peculiar and interesting Helix in Palestine.'

H. Ledereri, Pfr. Yebnah, between Gaza and Jaffa. Similarly marked as the last, but thoroughly depressed instead of trochiform.

H. Seetzeni, Koch. Wâdy 'Arabah; Sinai, at Wâdy Barak; abundant in Judæa, often covering the stems of Anabasis.

H. candidissima, Drap., var. minuta. Sinai, near 'Akabah; Wâdy Ghurundel, in Edom; Jordan; Bîr es Sebâ.

H. prophetarum, Bourg. Wâdy Ghurundel, Edom; Wâdy 'Arabah.

H. Boissieri, Charp. Judæa. A large solid species thinly distributed on rich land between Bîr es Sebâ and Gaza.

H. filia, Mouss. Wâdy 'Arabah, in several places; Wâdy Ghurundel, in Edom; Ghôr es Safieh. Canon Tristram found it 'extremely scarce, only in a few localities near the Dead Sea.'

H. spiriplana, Oliv. Jebel Abu Kosheibeh and Mount Hor; Wâdy Ghurundel, in Edom; and between Jericho and Marsaba. Jerusalem, at the ruins of St. John's Hospital. (= *H. guttata*, Bourg.)

H. cavata, Mouss. Bîr es Sebâ and Tell el Milh (Moladah); Gaza.

H. Olivieri, Feruss. Jericho, by the Jordan; Ghôr es Safieh, abundant dead amongst drift on the shore of the Dead Sea. Has not, I think, been gathered in Palestine before.

Helix bætica, Rossm. Drift on the shore of the Gulf of 'Akabah. Derived from the Midianitish side.

Bulimus cadmeus, Pfr. Dead Sea drift; Wâdy Ghurundel, and elsewhere in the 'Arabah to 'Akabah.

REPTILIA.

REPTILIA.

ORDER OPHIDIA.

The following have been determined by Dr. Gunther:

Rhyncocalamus melanocephalus, Gunther. This little snake was captured by Laurence at Petra. The genus, which was founded by Dr. Gunther for this species, contains as yet no other representative, and it was believed peculiar to the Jordan Valley. As in other instances, the Wâdy 'Arabah, or the wâdies leading to it, extends the range of the Jordanic forms of life. According to Wallace, with the exception of three American species, the small family of Oligodontidæ, to which this genus belongs, is confined to India, China, Japan, and the neighbouring islands.

Zamenis ventrimaculatus, Gray. I shot an individual of this species in Wâdy Zelegah, in Sinai. This snake ranges from Egypt to Beloochistan.

Z. atrovirens, Shaw. A snake of this species was killed on the plains of Southern Judæa. It is common in Southern Europe and Palestine. The black variety, *Z. carbonarius*, was, I believe, killed during our march in Sinai, in Wâdy Nasb, but it was unfortunately destroyed.

Z. elegantissimus, Gunther. This prettily-marked little snake was captured at 'Akabah. It was described first by Dr. Gunther from a specimen brought from Midian in lat. 26° 30' in 1878. Its distribution is,

Missing Page

Missing Page

therefore, so far as is known, confined to the eastern shores of the Red Sea.

Zamenis Cliffordii, Schleg. Slopes of Mount Hor. Known from Egypt, Algiers, and West Africa.

Cælopeltis lacertina, Wagl. Jebel Herteh, in Sinai. This snake is a native of North and West Africa, according to Wallace; North Africa, Arabia, and Persia, according to Gresham. It is the only one of the genus.

ORDER LACERTILIA.

Acanthodactylus boskianus, Dand. Sinai and 'Arabah.

Gongylus ocellatus, Forsk. Sinai, 'Arabah, and Ghôr; common.

Sphænops capistratus, Wagl. Wâdy Ghurundel, on the western side of Sinai. I found this skink hidden in an ant hill of Camponotus.

Ptyodactylus Hasselquistii, Schneid. I first caught this gecko in Wâdy Zelegah, in Sinai. It became frequent afterwards in the 'Arabah and the Ghôr.

Stenodactylus guttatus, Cuv. Sinai.

Agama sinaiticus, Heyden. Sinai, at low levels, and 'Akabah. A handsome blue-throated lizard.

Agama ruderatus, Oliv. Very common throughout Sinai and Wâdy 'Arabah to the Dead Sea.

Eremias guttata. 'Akabah, Sinai.

E. gutto-lineata. Ghôr.

Chameleo vulgaris, Dand. Sinai, at 'Ayûn Mûsa.

In the Amphibia, *Rana esculenta*, Linn., and *Bufo viridis*, Laur., were both brought home; the latter from Wâdy Ghuweir, and the former from the Ghôr. At Jericho, these frogs kept up an extraordinary din throughout the night. I may mention here that in the Ghôr, at the south end of the Dead Sea, abundance of land, or rather marsh, crabs occur of the species *Telphusa fluviatilis*. They are not confined to the water, for one was killed foraging in our camp, and I saw them amongst the rushes escaping with marvellous celerity. I brought home a number which the Bedawin children caught for me. They are now in the British Museum.

Tristram mentions that there were fresh-water crabs at 'Ain Jidy (Engedi); they were probably the same species. These crabs chiefly appear to resort to deep holes, which they excavate among the rushes above the margin of the Dead Sea. Here they lie concealed during the day, coming forth at dusk. When alarmed, they escape to concealment with great swiftness. This species is found in Egypt, North Africa, and eastwards to Persia.

AVES.

AVES.

THE following birds were observed in Sinai, Arabia Petræa, and South Palestine from November to February. Those whose names have a † prefixed were kindly identified for me by Canon Tristram, from specimens brought home. For fuller information the reader should consult Canon Tristram's work, and also Mr. Wyatt's appendix to the 'Ordnance Survey of Sinai.'

†*Turdus iliacus*, Linn. Redwing. I shot a redwing at 'Ain es Sultân, Jericho, on January 15. Not obtained, I believe, in Palestine previously.

†*T. merula*, Linn. Blackbird. I saw several, and shot one in the latter half of December in the Ghôr es Safieh. Blackbirds were seen also at Gaza in the following month.

†*Saxicola monacha*, Temn. Hooded chat. In Wâdy Nasb and Wâdy el Hamr in Sinai, I obtained this species, and again in the 'Arabah at 'Ain el Tâbâ, where there were several birds, mostly males. I also met it near Tell el Milh in the Judæan wilderness. Canon Tristram identified four specimens amongst my collection. This species is found elsewhere in Egypt and Nubia.

†*S. isabellina*, Rupp. Menetries wheatear. I obtained several specimens of this bird in Sinai, the 'Arabah, and near Bîr es Sebâ. It was frequently in company with the hooded chat. A Mesopotamian species ranging from North-east Africa to China.

S. mœsta, Licht. Tristram's chat. About four miles south-west of Bîr es Sebâ I fell in with a few chats I had not previously seen. They had a most peculiar call, a sort of exclamation in two strangely unlike notes, which differed from that of any other bird I have heard. They were very wary, and kept well out of range of Cairo gunpowder. They seemed to be about the size of the hooded chat, but presented more varied colouring, without white. From my notes, Canon Tristram refers these birds to this species. Found from the Sahara to Arabia.

Saxicola œnanthe, Linn. Wheatear. Ramleh and Jericho. A spring visitant to Sinai.

S. leucopygia, Brehm. White rumped chat. Frequently seen in Sinai, as in Wâdy es Sheikh and elsewhere, and in Wâdy Ghuweir in Edom. I shot several specimens. This bird has a very sweet, but weak, low, murmuring song, to which I often listened with pleasure. It is found elsewhere in the deserts of Sahara, Nubia, and Arabia. *S. leucocephala*, Brehm., a form of this species was also seen, I believe, at 'Akabah.

S. lugens, Licht. Pied chat. Met with in the desert and in the Ghôr, but more common about Jerusalem. *S. Finschii*, Hengl., was also seen about Jerusalem and at Bâb el Wâd. These two chats are both natives of Arabia, Egypt, and Algeria, but their headquarters are, according to Canon Tristram, in Palestine.

†*Cercomela melanura*, Temn. Desert blackstart. One of the frequent species all through Sinai to Ghôr, and at Jericho. Found in Arabia, Egypt, Nubia, and Abyssinia.

†*Pratincola rubicola*, Linn. Stone chat. This species was not unfrequent in the Ghôr es Safieh, near the Dead Sea. Mr. Holland met with it in Sinai, Ramleh, and Jericho.

†*Ruticilla tithys*, Linn. Black redstart. I obtained this bird at Jerusalem, where it was frequent, as also at Gaza. Mr. Wyatt found it common in the upper part of Sinai.

†*Cyanecula cærulcculus*, Pall. Blue-throated warbler. Frequent at 'Ayûn Mûsa, and again at 'Akabah, at both which places I obtained specimens. I shot it again in the Ghôr, where it was common. This species was very tame.

Erithacus rubecula, Linn. Robin. Between Ramleh and Jerusalem, and in the Ghôr at Jericho.

Sylvia nana, Hemp. et Ehr. Pigmy warbler. Wâdy Zelegah, northeastern escarpment of the Tîh, amongst tamarisk bushes; Wady 'Arabah at El Tâbâ ; Ghôr es Safieh. I obtained two specimens, one at Wâdy Zelegah, and one in the Ghôr. A desert species found in Sahara, Turkestan, Scinde, and South Persia.

†*Phylloscopus rufus*, Bechst. Chiff-chaff. Frequent at wells in Sinai ; once it uttered its summer note at 'Ayûn Mûsa, in November, in a broken but unmistakable fashion.

†*P. trochilus*, Linn. Willow wren. Also frequent throughout Sinai. Both these were often shot, and appeared to be the only warblers of any frequency.

†*Argya squamiceps*, Rupp. ' Hopping thrush.' From 'Akabah in several places wherever groves of acacia occurred along the 'Arabah to the Dead Sea. Abundant at the Ghôr es Safieh, and not unfrequent at Jericho. Some of these birds appeared to be mated in the beginning of December. On shooting one I saw, on two occasions, another perch on an acacia bush and utter a peculiarly piteous cry for a considerable time. But as they always go in small flocks, it may have been a purely social regret. Canon Tristram believed this bird to be confined to the Dead Sea, mentioning, however, that it was said to be found near 'Akabah, and in the Hedjaz. El Tâbâ is the best bird ground in the 'Arabah ; here there were many seen.

†*Drymœca gracilis*, Licht. Plentiful in the Ghôr es Safieh. Ranges from North-east Africa to Asia Minor and India.

D. inquieta, Rupp. Frequent throughout Sinai, chiefly on the eastern side, in low bushes of tamarisk in dry, stony wâdies, hopping, hiding, cocking its tail, and concealing itself with great dexterity. Confined to Sinai and Palestine.

Parus major, Linn. Ox-eye tit. Between Jerusalem and Jaffa.

Motacilla alba, Linn. White wagtail. From Alexandria and Cairo to the Ghôr and Jaffa this bird was frequently met with. In the desert of Sinai it is the commonest and the tamest species in winter, fearlessly hopping amongst the tents and camels.

M. flava, Linn. Yellow wagtail. Several of these were observed at 'Akabah along the shore. Mr. Wyatt records this species from the same locality as belonging to var. *cinereo-capillus*, Savi.

Motacilla sulphurea, Bechst. Gray wagtail. Not unfrequent in the Ghôr es Safieh. Mr. Wyatt met this bird in Sinai.

? *Anthus trivialis*, Linn. I believe this was the pipit which occurred in the Ghôr. Unfortunately my specimen was too bad for determination.

†*Pycnonotus xanthopygus*, Hemp. et Ehr. 'Akabah; El Tâbâ and Wâdy Ghuweir on the 'Arabah; Petra; very abundant in the Ghôr es Safieh; at Gaza and Jericho. In the Upper Ghôr el Feifeh every bush of salvadora (which forms dense groves) contained these birds. I have never seen any birds, not in flocks, so numerous. They had some soft, sweet, trilling notes, but were not, I presume, in proper song. This bird is found outside Palestine and Sinai in Northern Syria. Captain Burton met it to the southern limit of his travels in Midian in lat. 26° 30' ('Land of Midian,' ii. 185), so that it is perhaps tolerably wide-spread in Arabia. Mr. Wyatt found the bulbul in Wâdy Feiran in Sinai.

†*Lanius fallax*, Finsch. Frequent in Sinai, and up the 'Arabah to the Ghôr. Gaza. Found in Nubia, Abyssinia, and eastwards in Persia and Beloochistan.

Hirundo Savignii, Streph. In great abundance between Gaza and Ascalon, and about Medjel along the coast. For some miles these birds flew about and beside our horses, coming within arm's length, and displaying the showy under-colour, which is continued to the tail, and seems to be almost the only distinction from the common swallow. It seemed somewhat larger than *H. rustica*, of which a small number occurred intermixed. Mr. Sharpe informs me that he does not consider it distinct; but besides its colouration, it is non-migratory, and confined to Egypt and Palestine. Also seen in Sinai by Mr. Wyatt. It may perhaps, therefore, be regarded as only a climatic variety.

H. rustica, Linn. 'Akabah, on the confines of Sinai, and along the coast from Gaza to Jaffa.

Cotile rupestris, Scop. Ghôr es Safieh.

C. obsoleta, Cap. I saw this species at Wâdy Nasb, and also, I believe, in Wâdy el 'Ain. Its pale back and very square tail were distinctly noticed. It is a desert species, found in Nubia ; also Arabia and Persia.

†*Cinnyris Oseæ*, Bonap. Mr. Armstrong, who was with me at El Tâbâ, near 'Akabah, saw a pair of these birds amongst the acacia bushes so abundant along the east side of the 'Arabah near that place. I myself heard a note there which I afterwards recognised as belonging to this species at the Ghôr. It is interesting to extend the range of this bird nearly to 'Akabah, along the same valley southward as the hopping-thrush, and points out the direction perhaps by which the ancestral species originally spread to Palestine to habitats now almost isolated by changed circumstances. The sunbird delights in the long tubular flowers of the showy *Loranthus acaciæ*, which perhaps requires its visits for cross-fertilization. A specimen shot by Dr. Hull in the Ghôr es Safieh had its beak covered with the pollen. At El Tâbâ the loranthus is abundant, as also at the Ghôr; and at the latter place acacia trees, tenanted by this parasite, were more favoured by these birds, which were seen to probe their flowers. The sunbird was seen at Jericho ; and Dr. Port, of Beirût, informed me that they had visited his garden more than one year, and had on one occasion nested there. Beirût to 'Akabah would define its limited range, so far as yet known. Burton, however, saw sunbirds—almost certainly this species—scattered throughout Midian, especially in the Hisma country, in latitude 27°.

Carduelis elegans, Steph. Goldfinch. Frequent in the open country near Jerusalem, Gaza, and Jaffa.

†*Passer hispaniolensis*, Temn. Marsh-sparrow. First seen at Wâdy Hessi and 'Akabah. In large flocks along Wâdy 'Arabah and at the Ghôr.

P. domesticus, Linn. Common sparrow. Alexandria, Gaza, and other towns. In large flocks in the desert a day's ride east of Gaza.

P. moabiticus, Tristram. Tristram's sparrow. I saw this species on more than one occasion in small flocks close to the edge of the Dead Sea. I never succeeded in getting a shot ; but from its small size, I felt sure I was looking at this most local species, which I was familiar with from Canon Tristram's account in the 'Land of Israel.' It seemed fond of the saline and filthy mud, and hopped amongst the calcined tamarisk bushes on the outermost limits of vegetation below our camp. This Passer has not been found beyond the Ghôr.

Fringilla cælebs, Linn. Chaffinch. Gaza, and in the Ghôr at Jericho; also between Jerusalem and Bâb el Wâd. A few scattered birds. Mr. Wyatt obtained chaffinches in Sinai.

†*Carpodacus sinaiticus*, Licht. Sinaitic rose-finch. A single specimen of this beautiful little bird was shot by Dr. Hull from a flock of marsh-sparrows in the Wâdy 'Arabah. The rose-finch is confined to Sinai and South Palestine.

Erythospiza githaginea, Licht. I obtained this bird in Wâdy Nasb. Subsequently in Judæa, south of Bîr es Sebâ, I heard a curious, low, ventriloquizing note, rising and falling, and with a faint resemblance to our yellowhammer's cry, which I have reason to believe belonged to this species. I could not get a shot at it. My Bedawîn called it 'thyur.' A desert species, ranging from the Canaries to Scinde.

†*Emberiza miliaria*, Linn. Common bunting. Frequent amongst the dhoura fields in the Ghôr es Safieh, and in the Judæan wilderness, following the sowers.

†*E. striolata*, Licht. In the Wâdy 'Arabah, and at Wâdy Ghurundel in Sinai. A desert species, found from Morocco to North-west India.

Sturnus vulgaris, Linn. Starling. Between Jerusalem and Jericho in vast flocks.

S. unicolor, Temn. Sardinian starling. The 'black starling' came to roost in some numbers in the marsh at Tell Abu Hareireh the night we were encamped there. Subsequently I saw a few near Ramleh, at the western base of the Jerusalem plateau.

†*Amydrus Tristramii*, Sclater. Tristram's grakle. Small flocks of this species were met with at Jebel Abu Kosheibeh, and in the wâdy below it leading into the 'Arabah, about halfway between the Ghôr and 'Akabah. A few were seen at Petra, and a flock kept circling round the summit of Mount Hor all the time I was botanizing there. At first, while expecting to be molested by the Bedawîn, Dr. Hull and I believed this bird's clear whistle was a signal amongst the mountaineers. I obtained a specimen

at the Ghôr es Safieh. This bird is only known from one or two localities in Sinai, and from the Dead Sea neighbourhood. Its nearest allies are found in East Africa.

Garrulus atricapillus, Isid. East and west of Jerusalem. Frequent in olive trees.

Corvus cornix, Linn. Abundant about Gaza along the shore, and at Ascalon. Plentiful at Cairo.

C. affinis, Rupp. Fantail raven. Though I failed to obtain this species, I have no doubt it was the bird of Petra. It was very wild, and the clear musical cry mentioned by Canon Tristram had attracted my attention. I saw it first on the slopes of Mount Hor, about halfway between Petra and the summit. There were about half a dozen birds. This is an Abyssinian species, which has elsewhere been found only at the Dead Sea by Tristram, and the present link in its range is therefore important.

C. umbrinus, Hedenb. Frequent all through Sinai.

C. corax, Linn. Often seen in Palestine, and on one or two occasions I felt sure, from its superior size, that I saw it in Sinai.

C. frugilegus, Linn. (*C. agricola*, Trist. ?) I met a small flock of rooks, and shot one in the Ghôr.

†*Certhilauda alaudipes*, Desf. 'Persian lark.' I obtained this species first at Wâdy Hamr, afterwards on Jebel Herteh, and several times along the 'Arabah Wâdy. It uttered a low sweet song while on the ground. I saw it frequently in Sinai in the more northern part of the peninsula, and at 'Akabah. Ranges through sandy deserts from North Africa to Scinde.

Galerita cristata, Linn. Crested lark. I saw this species on both corners of the Sinaitic peninsula at Suez and 'Akabah. Also at the Ghôr, and in the Judæan wilderness.

†*Alauda isabellina*, Bonap. Isabelline lark. I met this species frequently in Sinai and the 'Arabah, and obtained at least half a dozen specimens. Confined to Sinaitic, Egyptian, and Sahara deserts, and extending like many other forms of life up the 'Arabah to the Dead Sea.

Alauda arvensis, Linn. (*A. cantarella*, Bp. ?) In large flocks in the Ghôr and the Judæan wilderness, following the sowers.

†*Ammomanes deserti*, Licht. Obtained frequently in Sinai and the 'Arabah, and perhaps the most constantly met with species in the barest desert. Occurred also between Tell el Milh and Bîr es Sebâ. Usually from two to five together, with a feeble, plaintive little song. A desert species ranging through the Sahara to Egypt, Nubia, and Abyssinia, and eastwards to Scinde.

Melanocorypha calandra, Linn. Calandra lark. Abundant between Gaza and Jaffa.

Caprimulgus tamaricis, Tristram. Several times I detected the unmistakable flight of a goat-sucker quickly disappearing amongst the nubk bushes in the Ghôr, but never succeeded in getting a shot. Canon Tristram's Survey enables me to state the species.

Picus syriacus, Hemp. et Ehr. Mr. Armstrong observed this species amongst the olive-trees a little north of Gaza.

Halcyon smyrnensis, Linn. Smyrna kingfisher. I saw this handsome species several times in the Ghôr; and on one occasion shot one, but could not discover it amongst reeds. Dr. Hull was more successful, and secured a splendid specimen. This kingfisher has an excessively loud and discordant, almost human cry, which it utters both at rest and on the wing. This species is Indian, ranging from China to Western Asia, and finding here its western limit.

†*Athene meridionalis*, Risso. Southern little owl. I saw this bird first at 'Akabah, where it was the object of persecution of several chats. Has been seen in Sinai. Subsequently at Gaza it was our constant companion amongst the sycamore trees in the quarantine ground. At Jericho also I both heard and saw it. Found in the Mediterranean countries and eastward to Afghanistan.

? *Bubo ascalaphus*, Savigny. A cry was heard on the wing in the wâdy leading to Es Sheikh from Dein el 'Arbain on Jebel Katharîna, which my Bedawîn called the hoōdoō. It was, I believe, that of an owl, and not unlike the deep coo-like note of the snowy owl, but more musical.

AVES.

The Arab believed it to be a bird, and I think it is probable it was this species.

Gyps fulvus, Gmel. Griffon vulture. This bird was, I believe, seen occasionally at considerable distances; but on one or two occasions in Wâdy Lebweh and in the 'Arabah, in lonely precipitous ravines, I came on it at unawares, and had full view of its noble proportions and stately flight. In Wâdy Harûn several gathered to the carcase of one of our camels.

Neophron percnopterus, Linn. Egyptian vulture. Hardly a day occurred on which this bird was not in view in our journey through Sinai, but especially on the African side of the peninsula.

Circus æruginosus, Linn. Frequent in the Ghôr.

Buteo ferox, Gmel. Saw one in the Ghôr. Has been obtained in Sinai by Mr. Holland.

Aquila chrysaetus, Linn. Golden eagle. From the summit of Jebel Abu Kosheibeh, one of the Mount Hor group, I disturbed a splendid pair of these eagles. They had permitted me to approach within a few yards, when, on looking over the cliff, they rose from immediately beneath me.

Milvus regalis, Linn. Kite. Frequent in Southern Palestine from Gaza along the coast, and in the Judæan wilderness.

Falco tinnunculus, Linn. Kestrel. The kestrel was first seen in Wâdy Ghuweir on the east side of the 'Arabah wâdy, between Petra and the Ghôr. At the Ghôr it was very frequent.

Accipiter nisus, Linn. Sparrowhawk. Not unfrequent in the Ghôr. A single bird of this species was the only living inhabitant on the summit of Jebel Usdum which I actually saw. Feathers of a dove were, however, close by, where he had probably struck his prey.

Pelecanus crispus, Bruch. Dalmatian pelican. A pelican gave me capital ball-practice in a muddy estuary about four miles south of Gaza; but it only afforded me another of repeated instances of the utter uselessness of firing ball-cartridge from a fowling-piece, especially if backed by Cairo gunpowder.

Ardeola russata, Wagl. This species was seen several times by the railway between Cairo and Suez, where it was pointed out to me as the 'Ibis.' I met with it again at 'Ayûn Mûsa at the north-western extremity of the Sinaitic peninsula.

Ciconia alba, Bech. White stork. I saw a single stork on the shore below 'Ayûn Mûsa at the north-western end of the Sinai peninsula. Later on a small party remained in sight for some time near Bîr es Sebâ.

Anas crecca, Linn. Teal. A small flock of teal was sprung by me at Tell Abu Hareireh from a marshy pond. Mr. Wyatt has obtained teal in Sinai.

Fuligula ferina, Linn. Pochard. On two occasions I disturbed these birds while botanizing round the margin of the Dead Sea. They did not rise actually from the water, but from a deep stagnant salt pool about a hundred yards from its margin.

Columba Schimperi, Bp. The rock pigeon (I presume this form, for it never came within range) was seen first in the caves of Mokkattam, near Cairo. Subsequently we put several to flight from the cliffs of Wâdy el 'Ain; and I met with it again between Jericho and 'Ain Sultân.

Turtur risorius, Linn. Collared turtle-dove. Very abundant in the Ghôr es Safieh, and afforded us continual sport and capital eating. At roosting time we were generally able to procure a few brace amongst the nubk-trees (*Zizyphus spina-christi*). An Asiatic bird outside Palestine.

†*T. senegalensis*, Linn. Palm turtle-dove. This beautifully tinted bird was not so common as the last species, although plentiful. My experience of their relative abundance differs from that of Canon Tristram.

†*Pterocles senegalus*, Linn. Pintail sand-grouse. This species was first met with at 'Ayûn Buweirdeh, about a day's march south along the 'Arabah from the Ghôr. Afterwards it was frequent about Bîr es Sebâ in Judæa. Its odd cry, uttered on the wing, half coo, half croak, reminded me of that of the Manx shearwater. An African bird outside Palestine. Found in the deserts of Africa, Arabia, and Mesopotamia.

Caccabis chukar, G. R. Gray. Seen in the Wâdy 'Arabah and in some numbers south of Bîr es Sebá in Judæa. A bird of this species was tame about and in the house at the resting-place at Stora between Beirût and Baalbeck.

C. Heyi, Temm. Hey's partridge. First met with in Wâdy Nasb, and thence throughout Sinai to the Ghôr. On the west side of the 'Arabah it was especially abundant. Rather dry eating, and much smaller than the last species. Confined to Nubia, Egypt, and Sinai, and reaching the Dead Sea by the 'Arabah.

Coturnix communis, Bosm. Quails were seen by me in the Ghôr. Mr. Holland obtained quails in Sinai.

Rallus aquaticus, Linn. Water rail. One was obtained close to the Dead Sea in a salt pool.

Grus communis, Bechst. Crane. At Tell el Milh (Moladah) a considerable flock of cranes was met with; they were extremely shy. For the next two marches, close to Gaza, their trumpeting was not unfrequently heard at night. Mr. Wyatt found cranes constantly at Tôr in Sinai.

Œdicnemus scolopax, Gmel. Thick-knee. One of these birds was brought to me by an Arab, who had shot it close to the Ghôr in the 'Arabah.

†*Ægialites asiatica*, Pall. Caspian plover. Several of these, of which one was shot, was feeding along the shores of the Gulf of 'Akabah.

Vanellus cristatus, Meyer. Lapwing. Single birds were seen several times in the Ghôr; large flocks occurred on the Judæan plain.

Gallinago media, Leach. Snipe. Several were seen in the Ghôr.

Rhynchæa capensis, Linn. Painted snipe. From the notes I took on the spot and the descriptions and figures I have since consulted, I have no hesitation in referring to this species the snipes occurring in the salt swamps at Tell Abu Hareireh, a few miles east from Gaza. There were three or four birds, but unfortunately I failed to shoot one, although I obtained the chance. One would rise from the rushes and, after a short flight near the ground, pitch himself into the mud again and commence a peculiar quacking cry. The colouring, which looked very gay with green, brown, and white, seen from above on the wing, was observed closely, and corresponded unmistakably with this species. This bird is known throughout Egypt to Nubia, but has not been previously noticed in Palestine.

Tringa alpina, Linn. Dunlin. Coast at Gaza and 'Ayûn Mûsa.

†*Tringoides hypoleucus*, Linn. Sandpiper. At 'Akabah and on the shores of the Dead Sea.

Totanus ochropus, Linn. Green sandpiper. Dead Sea shore. This species and the last, in small parties, ran wading along the muddy edge of the Salt Sea at the swampy ground caused by the Feifeh stream. They waded right into the salt water, where I followed and shot them. Mr. Wyatt met these birds at Tôr in Sinai.

T. calidris, Linn. Seen several times in the Ghôr and at Tell Abu Hareireh.

Sterna minuta, Linn. I saw this species at Port Said and again near Gaza.

Larus ridibundus, Linn. A few were met with in the Judæan desert, where they forage after the multitudinous helices found there. Seen at Tôr by Mr. Wyatt.

L. minutus, L. 'Akabah.

The foregoing account will serve to show the species likely to be met with in the winter months, the time usually selected for an Eastern tour of this description. Several other species which could not be identified were met with, and perhaps it may be thought I have included too many species which were not actually obtained, but of these many were British and familiar birds, and the rest usually unmistakable. Moreover, many birds were obtained which were not brought home, sometimes owing to their bulk, but usually for the reason which eventually lost me all my specimens, their not being skinned. I had no one to help me in this respect, and I found it hard enough to get requisite sleep after each day's work, when my botanical collections were sorted, without attending sufficiently to my other gatherings. I depended, therefore, on carbolic acid, which I used without extracting the entrails, and the result was unfortunate. In connection with the ornithology of this district, a valuable paper by Mr. Wyatt, appended to the 'Ordnance Survey of Sinai, 1869,' should be consulted, and, taken in connection with the great work of Canon Tristram, leaves little to be discovered. There are, however, parts of Sinai hardly visited even still, and the Wâdy 'Arabah, upwards of a hundred miles in length and hardly known to naturalists, has not yet been explored at the best season. This wâdy is the most interesting ground by far, except, perhaps, the Ghôr es Safieh. All the desert species of Sinai may, I believe, be obtained there, and several of the most interesting Dead Sea species. Desert larks and chats, of about ten species, occurred in this wâdy. Hey's partridge, Tristram's grakle, Palestine bulbuls, hopping thrushes, sunbirds, fantail ravens, rose-finches, and others, some of them of extreme rarity and interest in geographical distribution, were observed either in this valley or some of its system close by. As with the plants, so with the birds, this valley has afforded a low-lying highway, easily traversed in bygone times when the climate was moister and vegetation more luxuriant, from Sinai, Egypt and the hill valley on the one side; from the Hedjaz and Happy Araby on the other, to the rich oases around the Dead Sea. Their descendants—sometimes modified, sometimes unchanged, and sometimes there alone surviving—still remain in favoured spots, and chiefly in the southern Ghôr. The rodents tell the same history, a history which still stands in need of much detail before it can be thoroughly dealt with.

About a third of the above list of species are British, chiefly summer migrants here, and occurring in Sinai and Palestine as winter visitants. The remainder of the Sinai birds will include the most interesting portion of the group, the *desert* birds of Sinai, and of these I give a separate list. They are probably all residents. One or two which have been obtained only on the confines of Sinai are included in brackets. Those few which are not mentioned in the foregoing account are taken from Mr. Wyatt's list above mentioned. The total number of species of birds which visit the peninsula may be probably estimated at a hundred.

Saxicola monacha, Temm.

S. deserti, Temm.

S. leucopygia, Brehm. (et *S. leucocephala*, Brehm.).

S. lugens, Licht.

Missing Page

Missing Page

Aucs acuta, Linn.
Mergus serrator, Linn.
Phalacrocorax carbo, Linn.
Larus gelastes, Licht.
L. melanocephalus, Natt.

MAMMALIA.

MAMMALIA.

Hyrax syriacus, Hemp. et Ehr. Coney. Seen only on the summit of Jebel Mûsa, but occurs in warrens amongst boulders in several places amongst the Sinaitic mountains. This peculiar little animal, which belongs to the only genus of the order called after it, and placed next the elephants, is the sole representative found outside the African continent, where two or three other species of hyrax are found. It is the *shaphan*, or 'coney,' of the Bible, the *daman*, or *wabar*, of the Arabs, and probably the *ashkoko* of Ethiopia. See Cassell's Nat. Hist., ii. 292; article by Messrs. Boyd Dawkins and Oakley. See also Tristram's Nat. Hist. of Bible, and Wyatt's Ord. Survey of Sinai, App. For earlier notices, see Laborde's 'Arabia Petræa'; and for the views of ancient writers, and an excellent general account, I would refer the reader to a too little known work—Harris's 'Natural History of the Bible,' 1824. The coney is found from Ethiopia and Arabia to Arabia Petræa and Lebanon. Captain Burton obtained it in Midian on Jebel el Shan, latitude 27° 30'.

Sus scrofa, Linn. Boars were actually seen by our party only near Bir es Sebá, where their fresh traces were also very abundant. In the reedy jungles south of the Dead Sea, although I heard these animals crashing in front of me once or twice, and came frequently on the 'spoor' of old and young in the mud, which I followed into the thickest parts, I never obtained a shot. In the bare Judæan plain between the Ghôr and Gaza, where there is rarely cover or hiding-place, these animals abound on account of the plentiful supply of bulbs on which they feed. Bulbs of *Urginea* (Scilla), and especially *U. undulata*, Desf., are much esteemed ;

and of the latter I found it difficult to obtain a root which was not mashed. No traces of boars were seen in the Wâdy 'Arabah. The Palestine wild boar belongs to the European, and not to the Indian, form. Found throughout Europe to Syria.

Gazella dorcas, Linn. Gazelles were not seen in Sinai, but their tracks were occasionally met with. They did not abound, however, till reaching Wâdy Zelegah, and other wâdies skirting the Tîh plateau. In Wâdy 'Arabah gazelles were frequently seen in parties of from three to seven, and chiefly on the western side. On the Judæan plain from Bîr es Sebâ to Gaza these graceful animals were much more common, wilder, and in larger herds. At Jerusalem I saw one which was killed during the snow within a mile of the Holy City in the middle of January. They very rarely occur so close to the town. The Dorcas gazelle is found in Syria, Arabia, Egypt, and Algiers.

Capra beden, Wagn. Ibex. Good heads of the Syrian ibex were obtained from the Arabs in different parts of the peninsula, but it was not until reaching the Wâdy 'Arabah that the animal was met with alive. They were first seen on the plateau below the summit of Mount Hor, and again on the verge of the marls at the south-east of the Ghôr, about 600 feet above the level of the Dead Sea. In the latter place I had an excellent view of three at about a hundred yards' distance. They uttered a whistling snort like that of Highland sheep when they perceived me, and went off at by no means a rapid pace, although I urged them on with a round of shots, including ball-cartridge and 'swandrop.' The 'beden' is found in Egypt and Arabia Petræa. No other wild ruminants, whether sheep or antelope, could be heard of from the Arabs, either of Sinai or 'Akabah, or along the 'Arabah. They say their goats are an indigenous race, but that their sheep come from eastwards.

Lepus sinaiticus, Hemp. et Ehr. Sinaitic hare. First seen in Wâdy Berrâh, and easily recognised by its long ears, legs, and small body. Along the eastern Tîh escarpment it was often met with, and obtained in the Wâdy 'Arabah. It appeared to me to be possessed of greater speed than any other hare I have met with, and travelled at an altogether marvellous rate. Nevertheless, it is very stupid, and allows itself (like the Arctic hare) to be followed, and obtained by the clumsy weapons of the natives. After its first flight it appears, when again approached, to regard its enemy with less concern, and thus enables the Arabs to secure it, who usually require several discharges of their musket at short distances and stationary objects. Apparently confined to Sinai and the 'Arabah, but will probably be found in Arabia.

Lepus ægyptius, Geoffr. Egyptian hare. This species was not unfrequent on both sides of the 'Arabah, where the embouchures of mountain wâdies afforded more vegetation. It was seen at Wâdy Harûn, and on the declivities abreast of Wâdy Harûn, on the opposite side of the 'Arabah, in frequency. The Egyptian hare approaches our own hare in size.

Hystrix cristata, Linn. Porcupine. Quills were picked up between the 'Arabah watershed and the Dead Sea in several places, and at the Dead Sea at both the northern and southern extremities. Captain Burton obtained similar evidence of the porcupine, commonly in Midian on the east side of the Red Sea to latitude 26° 30'. These were probably the same species, but Canon Tristram thinks the Asiatic species (*H. hirsutirostris*) may occur. It is stated to have been obtained near Jerusalem. The porcupine is found throughout the Mediterranean region to Syria and Western Arabia.

Acomys, sp. ? Porcupine mice. Species of this desert genus of rodents appear to abound in the 'Arabah, but none were obtained. Three forms—*A. dimidiatus*, Rupp.; *A. cahirrhinus*, Geoff.; and *A. russatus*, Wagn.—have been taken in Sinai, all of which extend along the 'Arabah to the Ghôr. They are very abundant in some of these forms in the 'Arabah. Even on the barren summit of Jebel Usdum, a most desolate and parched eminence of salt, chalk, and dust, I found their burrows. In Sinai these were also observed in the higher wâdies, but in no such abundance. The first of these species is found in Nubia, Egypt, and Arabia Petræa; the second is Egyptian. Mr. Wyatt obtained these three mice about the convent of Mount Sinai, where also Mr. Holland obtained a species of dormouse, *Myoxus quercinus*.

Mus bactrianus, Blyth. Sandy mouse. This species is very abundant in the Ghôr es Safieh. A wire-trap in my tent at night caught specimens almost as fast as it was set. The Arab village or encampment of the Ghawarhineh tribe was close by, which accounts for their abundance. My specimens were determined by Mr. Thomas of the British Museum.

Gerbillus erythrurus, Gr. Desert rat. I trapped this species in the Wâdy 'Arabah at our camp on December 11th, south of 'Ayûn Buweirdeh. Mr. Thomas, who named my specimens, informs me that it ranges from Candahar, through Persia, Asia Minor, and Arabia to Algiers. It does not appear to have been taken before in Palestine or Sinai, and is not in Canon Tristram's fauna. This species (or *G. pygargus* ?) was also seen in Sinai, where its holes and tracks are not unfrequent. In the 'Arabah the burrows of the gerbilles and porcupine mice, forming centres of innumerable radiating lines of tracks, and often surrounded and partially filled with chopped vegetable matter, are most notable.

Psammomys obesus, Rupp. A specimen of this rat was obtained near Bîr es Sebâ, and identified by Mr. Thomas. Canon Tristram has noticed its abundance in the plains of South Judæa. Its burrows, and not very rarely the animal, are everywhere to be seen from Tell el Milh to Gaza. It is found also in the deserts of Northern Africa.

Spalax typhlus, Pall. Mole-rat. This animal is abundant from Bîr es Sebâ westwards, about Gaza, and for several miles south of that place. The mole-rat is found in South-eastern Europe and Asia as far as the Caucasus, and appears somewhat unexpectedly so far south as Gaza at sea-level. My specimens from Judæa are now in the British Museum.

†*Dipus ægyptius*, Licht. Jerboa. The jerboa was seen several times in the Wâdy 'Arabah, always just disappearing, and usually at the close of daylight. This species is found in North African deserts and south to Nubia; also in Arabia and South-western Asia generally.

Felis pardus, Linn. Leopard. Fresh tracks of the leopard or 'nimr' of the Arabs were met with at 'Ain el Tâbâ, a little north of 'Akabah. This spring is a favourite watering-place for beasts of various kinds. At

'Ayûn Buweirdeh, farther north in the 'Arabah, Mr. Laurence came upon fresh remains of some animal, with plentiful tracks of this carnivore around it. The Sinai Bedawîn stated that the 'nimr' is found only on Jebel Serbâl and the Shaumer.

Felis maniculata, Rupp. Mr. Laurence shot a fine specimen of this wild cat near Tell Abu Hareireh in Southern Judæa, between Gaza and Bîr es Sebâ. Our Cairo dragoman pronounced it to be the 'Nile wild cat.' It measured from tip of nose to tip of tail 2 feet 8 inches, of which the tail measured 1 foot. The animal was grayish brown above, faintly brindled across the back and down the sides with fulvous brown, the belly fawn-colour and barred or mottled with sooty black. The tip of the tail was black, with two or three rings of black near the tip. This is the cat found embalmed in Egyptian monuments. It ranges through Egypt and up the Nile to Abyssinia.

Hyæna striata, Schreb. On the western side of the 'Arabah the striped hyæna appears to be plentiful. I had a shot at one here, and saw their tracks in several places. In Sinai hyæna tracks were seen and recognised by the Arabs frequently.

Canis aureus, Linn. Very abundant in the Ghôr and at Gaza, but not heard or seen in Sinai. Sometimes the dogs seen along the coast from Gaza to Jaffa are strongly tainted with jackal. The bushy hanging tail and slouching gait, with pointed ears and long snout, indicated the parentage of these animals; and I was informed that the village dogs and jackals frequently interbreed. On Lebanon, in deep snow, between 4,000 and 5,000 feet above sea-level, I saw jackals with many ravens gorging on the carcase of a dead camel. Here also I saw the only wolf (*Canis lupus*) met with during the tour, though tracks were said to belong to this species in Wâdy Harûn at the base of Mount Hor. Burton saw a wolf in Midian.

Vulpes nilotica, Rupp. ? Fox-burrows were met with in the 'Arabah, probably belonging to this species. The Sinaitic Bedawîn were acquainted with foxes in Sinai. In the Ordnance Survey two other species of fox (*Canis fameticus*, Rupp. and *C. sabbar*, Ehr.) are given which were trapped in Sinai. The former of these is distinct from the true 'fennec fox' (*C. zerda*, Gm.) known all over Africa. It is the fennec fox of Nubia described by Ruppell.

Erinaceus brachydactylus, Wagn. Hedgehog. Twice in the 'Arabah I picked up the dried 'pelts' of this animal. This species is common in Palestine and Southern Egypt; it is no doubt the hedgehog found in Midian by Captain Burton on the east side of the Red Sea in latitude 26° 30'. It is common in Palestine and Egypt, and most likely occurs in Sinai.

INDEX.

A

ABUTILON fruticosum, 23, 33, 86, 131, 159
„ muticum, 86, 131, 159
Abyssinian plants, Nubian and, 150
Acacia, 58
„ albida, 64, 72, 92
„ balanites, 50
„ groves of, 153
„ nilotica, 9, 133
„ læta, 9, 52, 56, 92, 160, 164
„ salvadora, 50
„ seyal, 5, 6, 9, 12, 15, 24, 26, 35, 92, 133, 160
Acacia tortilis, 8, 9, 12, 28, 29, 30, 52, 79, 92, 133
Acacia zizyphus, 50
Acanthodactylus boskianus, 210
Acanthodium, 24, 25
„ spicatum, 14, 19, 25
Accipiter nisus, 223
Achatina (cionella) brondeli, 53, 204
Achillea santolina, 59, 134, 196
Acis elevata, 179
Acomys, 235
„ cahirrhinus, 235
„ dimidiatus, 235
„ russatus, 235
Acridium tataricum, 183
Adesmia, 5, 15
„ carinata, 179
„ lacunosa, 179
„ tenebrosa, 179
Adiantum capillus-veneris, 18, 118, 142
Ægialitis asiatica, 29, 225
„ fluviatilis, 229
Ægopyrum squarrosum, 142
Æluropus littoralis, 142
Ærua, 6, 22, 58
„ javanica, 6, 27, 107, 139
Agama ruderata, 8, 15
„ ruderatus, 210

Agama sinaitica, 13
„ sinaiticus, 210
Agrotis, 182
Agrostis, verticillata, 68, 116, 141, 167
Ainsworthia trachycarpa, 94
Aizoon canariense, 93, 133
„ hispanicum, 73, 93
Ajuga chia, 105, 138
„ iva, 59, 105
„ tridactylites, 39
Akadi, 28
Alauda arvensis (A. cantarella), 222
„ isabellina, 10, 15, 221, 228
Albersia blitum, 107, 139
„ caudatus, 107
Alcedo ispida, 229
Alhagi maurorum, 5, 16, 22, 62, 91, 133
Alkanna orientalis, 17, 18, 101, 136
„ tinctoria, 68, 101
Allium sinaiticum, 13, 39, 44, 113, 140
„ stamineum, 140
Alsine Meyeri, 130
„ picta, 130
Althæa rosea, 131
„ striata, 131
Alyssum campesire, 81
„ homalocarpum, 128
„ libyca, 62
„ (Koniga) libyca, 81
„ marginatum, 128
Amarantus caudatus, 138
„ sylvestris, 12, 139
Amberboa crupinoides, 97, 135
„ Lippii, 73, 97, 135
Ambystegium filicinum, 170
„ serpens, 170
Ammi majus, 73, 94
Ammomanes deserti, 10, 222, 228
Ammoperdix Heyi, 228
Amydrus Tristramii, 220, 228
Anabasis, 7, 17, 22, 35

Anabasis aphylla, 107
" articulata, 5, 7, 9, 10, 17, 18, 35, 62, 138
Anabasis setifera, 6, 9, 16, 17, 18, 22, 26, 32, 107, 138, 166
Anagallis arvenis, 99, 135
" latifolia, 99
Anagyris fœtida, 63, 64, 68, 89
Analysis of the flora of Sinai, 123
" orders represented in the Sinaitic Peninsula, 143, 144
Anarrhinum, 20
" pubescens, 14, 17, 102, 137
Anas erecca, 224
Anastatica hierochuntina, 25, 26, 33, 44, 81, 128
Anchusa ægyptiaca, 64, 101
" aggregata, 101
" hispida, 136
" Milleri, 59, 68, 101, 136
Ancillaria striolata, 194
Andrachne, 29, 32
" aspera, 14, 64, 110, 139, 162
Androctonus (Linrus) quinque-striatus, 17
Androcymbium (Erythrostictus) palæstinum, 44, 70, 112
Andropogon annulatus, 116
" distachyus, 141
" foveolatus, 29, 116, 141
" hirtus, 39, 63, 116, 141
Anemone coronaria, 62, 73, 79
Anisothecium (Dicranum) varium, 118, 168
Anthemis, 96
" deserti, 134
" kahirica, 134
Anthoceras lavis, 171
Anthrophora nidulans, 180
Anthus aquaticus, 229
" campestris, 229
" trivialis, 218
Antirrhinum orontium, 103
Anvillæa garcini, 34, 95, 134
Aphænogaster structor, 180
Apium graveolens, 133
Aquila Bonellii, 228
" chrysaetus, 223
Arabian flora, advance of, 151, 153
Arachnid (Sparacis), 25
Arbutus andrachne, 67, 99
Arca, 29
" antiquata, 201
" barbata, 193, 197, 200
" divaricata, 193
" fusca, 195
" retusa, 195
" scapha, 193

Ardeola russata, 4, 224
Arenaria graveolens, 17, 19, 20, 130
" picta, 73, 84
Argas persicus, 177
Argya squamiceps, 29, 31, 72, 182, 217, 228
Argynnys, 16
Argyrolobium uniflorum, 132
Arisarum vulgare, 111
Aristida acutiflora, 141
" ciliata, 32, 44, 116, 141
" cærulescens, 25, 116, 141
" hirtigluma, 141
" obtusa, 6, 11, 116, 141
" plumosa, 11, 111, 140, 141
" pumila, 141
" pungens, 141
Arnebia cornuta, 71, 101, 136
" hispidissima, 136
" linearifolia, 58, 136
Artemis rubecundus, 194
Artemisia, 22
" herba-alba, 14, 15, 19, 22, 62, 97, 134
Artemisia judaica, 5, 16, 134
" monosperma, 29, 44, 64, 97, 134
" santolina, 7
Arthrocnemum glaucum, 138
Artocarpus integrifolia, 69
Arum, 37, 39
" dioscorides, 68, 111
Arundo donax, 9
" phragmites (P. gigantea), 51
Asclepiad, boucerosia, 34
" pentatropus, 34
Asparagus acutifolius, 37, 65, 114
" aphyllus, 34, 37, 65, 68, 114
" stipularis, 140
Asperugo procumbens, 71, 102, 136
Asperula sinaica, 18
Asphodelus fistulosa, 37
" fistulosis, 14
" fistulosus, 113
" pendulinus, 140
" ramosus, 34, 35, 60, 62, 113
" tenuifolius, 140
Asplenium ceterach, 39
Asterella hemisphærica, 171
Astericus, 32
Asteriscus graveolens, 84, 95, 134
" pygmæus, 95, 134
" Schimperi, 134
Astragal (A. echinus), 20
Astragalus acinaciferus, 44, 90, 132, 164
" aleppicus, 64, 91
" alexandrinus, 59, 90

INDEX.

Astragalus bombycinus, 132
" callichrous, 71, 90
" echinus, 132
" eremophilus, 132
" Forskahlii, 44, 91, 132
" Fresenii, 132
" macrocarpus, 64, 90
" peregrinus, 132
" pseudostella, 132
" sanctus, 59, 71, 91
" Schimperi, 132
" Sieberi, 11, 91, 132
" sparsus, 132
" tenuirugis, 132
" tribuloides, 132
" trigonus, 11, 91, 132
Ateuchus sacer, 178
Athene meridionalis, 4, 222
Atractylis flava, 26, 97, 135
" prolifera, 64, 97
Atraphaxis spinosa, 139
Atriplex alexandrina, 58, 106
" alexandrinus, 166
" crystallina, 29
" crystallinum, 106, 138
" dimorphostegium, 138
" farinosum, 138, 162
" halimus, 9, 29, 106, 138
" hastatum, 106
" leucoclada, 9, 16, 23, 29, 51
" leucocladum, 106, 138, 166
" patulum, 106
" tataricum, 106
Atriplices, 26, 52
Atys (Alicula) cylindrica, 193
" " succisæ, 193
Aucs acuta, 230
Authors quoted, 198
Avena barbata, 141
" sterilis, 68, 116, 141
Avicula ala-perdicis, 195

B

Balanites ægyptiaca, 72, 88, 159
Ballota Schimperiana, 22, 23
" undulata, 14, 33, 58, 105, 137
Bapleurum linearifolium, 20
Barbula cuneifolia, 171
" lævipila, 171
" muralis, 171
" subulata, 171
" unguiculata, 171
" vinealis, 171
Batrachians, 55
Bellevallia flexuosa, 34, 37, 60, 113

Bellis sylvestris, 67, 97
Beta maritima, 138
Biscutella, 71
" columnæ, 59, 81
Blepharis edulis, 103, 137
Blumea (Erigeron) Bovei, 23
Boerhavia plumbaginea, 29, 108, 139
" repens, 53, 108, 166
" verticillata, 32, 35, 43, 53, 64, 108, 139, 162, 166
Boissiera bromoides, 142
Bollata Schimperi, 105
Botany of Central Arabia, 155
Boucerosia, 66
" Aaronis, 100, 165
" sinaica, 136
" sp. nov., 34, 38
Brachpodium distachyum, 142
Brassica deflexa, 82
" nigra, 82
" Tournefortii, 64, 82, 129
Brocchia cinerea, 134
Bromus madritensis, 73, 117, 142
" tectorium, 142
Bryonia multiflora, 93
" syriaca, 37, 39, 68, 93
Bryum argenteum, 119, 170
" atropurpureum, 59, 119, 170
" turbinatum, 20, 119, 170, 171
Bubas bubalus, 178
Bubo ascalaphus, 21, 222
Buccinulus solidulus, 193
Buffonia multiceps, 17, 19, 20, 84, 130
Bufo viridis, 43, 210
Bulimus, 53
" cadmeus, 203
" carneus, 41
" halepensis, 204
" Harti, 204
Bulla ampulla, 193
" striata, 197
Bupleurum linearifolium, 62, 94, 133
Buteo ferox, 223

C

Caccabis chukar, 225
" Heyi, 13, 225
Calandrella deserti, 228
Calendula arvensis, 59, 97, 135
Calligonum comosum, 29, 33, 43, 108, 139
Callipeltis cucullaria, 134
Caloptenus pilipes, 183, 185
Calotropis procera, 50, 72, 99, 135, 161
Calycotome villosa, 63, 65, 68, 69
Campanotus compressa, 9, 180

31

Campanula dulcis, 135
Camponotus pubescens, 8
Canis aureus, 237
　„　famelicus, 237
　„　lupus, 237
　„　sabbar, 237
　„　zerda, 237
Capparis galeata, 13, 18, 21, 22, 23, 25, 83, 129
Capparis Sodada, 159
　„　spinosa, 22, 39, 43, 82, 129
Capra beden, 234
Caprimulgus tamaricis, 222
Capsella, 71
　„　byrsa-pastoris, 62, 65, 71, 81
　„　procumbens, 71, 81
Cardita antiquata, 194
　„　variegata, 194
Cardium fragum, 194
　„　leucostoma, 201
Carduelis elegans, 219
Carduus argentatus, 68, 97
　„　pycnocephalus, 135
Carex, 39
　„　distans, 140
　„　divisa, 115, 140
　„　stenophylla, 37, 115
Carpodacus sinaiticus, 220, 228
Carrichtera, 71
　„　vellæ, 59, 82
Carthamus glauca, 34
　„　glaucus, 44, 62, 98
　„　lanatus, 34, 98
Carum, 94
Cassia acutifolia, 29, 91, 133
　„　bicapsularis, 3
　„　fistula, 28
　„　obovata, 29. 91, 133, 160
Cassis saburon, 191, 199
　„　torquata, 191
Caucalis leptophylla, 94
Caylusea, 22
　„　canescens, 14, 62, 83, 129, 163
Celsia, 20
　„　parviflora, 18, 20, 37, 102, 137, 165
Centaurea ægyptiaca, 135
　„　araneosa, 64, 97
　„　eryngoides, 17, 98, 135
　„　iberica, 97
　„　pallescens, 64, 97
　„　scoparia, 18, 33, 98
　„　sinaica, 97, 135
Cephalaria syriaca, 134
Cerastium glomeratum, 68, 84
Ceratonia, 65
　„　siliqua, 64, 91

Ceratophyllum demersum, 4, 85
Cercomela melanura, 12, 29, 72, 216, 228
Cerithiopsis clathrata, 192
Cerithium, 29
　„　cæruleum, 194
　„　Caillandi, 194
　„　echinatum, 192
　„　fenestratum, 192
　„　Kochi, 192
　„　lacteum, 192
　„　levantinum, 196
　„　mediterraneum, 196
　„　moniliferum, 192
　„　nodulosum, 201
　„　rostratum, 192
　„　scabrum, 196
　„　tuberosum, 192
Certhilauda alaudipes, 8, 221, 228
　„　desertorum, 10
Ceterach, 39, 67
　„　officinarum, 118
Chætoscladium trichospermum, 94
Chama, 29
　„　cornucopia, 193
　„　corrugata, 195
　„　iostoma, 195
　„　obliquata, 193
Chamæleo vulgaris, 4, 210
Chara hispida, 25, 118, 168
Charadrius Leschenalii, 229
Cheilanthes, 39, 67
　„　fragrans, 118, 142
　„　odora, 34
Chenolea arabica, 138
Chenopodium album, 106, 138
　„　murale, 106, 138
Chrysanthemum coronarium, 71, 96, 134
Cichorium, 71
　„　intybus, 98
Ciconia alba, 224
Cinereo-capillus, 217
Cinnyris Oseæ, 31, 219, 228
Circe abbreviata, 195
　„　arabica, 194
　„　crocea, 195
　„　lentiginosa, 194
　„　pectinata, 195
　„　tumefacta, 194
Circus æruginosus, 223
Cistus villosus, 83
Citrullus, 6
　„　colocynthis, 32, 62, 93, 133
Cladonia pyxidata, 120, 172
Clanculus Pharaonis, 192
Clematis cirrhosa, 68, 79

INDEX.

Cleome arabica, 7, 10, 14, 22, 82, 129
" brachycarpa, 129
" chrysantha, 129
" droserifolia, 23, 27, 82, 129
" trinervia, 14, 82, 129
Climate of Sinai, Professor Hull on the, 150
Clypeola microcarpa, 128
Cocculus Leæba, 32, 52, 79, 128, 159, 163
Cœsalpinia Gilliesii, 69
Cœlopeltis lacertina, 210
Cœlorachis hirsutus, 116
Colchicum, 61
" distans, 20
" montanum, 37, 70, 112, 140
" Ritchii, 140
" Steveni, 20, 37, 112, 140
Columba Schimperi, 24, 224
Columbella albina, 191
" albinodulosa, 191
" conspersa, 190
" ligula, 190
" (Ricinula) mendicaria, 190
" rustica, 195
Colutea, 39
" aleppica, 38, 90, 132, 164
Cometes abyssinica, 130
Compositæ, 19
Conus, 29
" arenatus, 191
" ceylanensis, 200
" generalis, 191
" hebræus, 200
" mediterraneus, 195
" monile, 195
" mussatella, 191
" mustellinus, 191
" nanus, 191
" nigropunctatus, 191
" pennacens, 191
" quercinus, 194
" tessellatus, 191
" textile, 191
" vicarius, 191
Convolvulus arvensis, 63, 101, 136
" hystrix, 136
" lanatus, 11, 100, 136
" siculus, 73, 100
Coralliophila (Rhizochilus) costularis, 190
" " exarata, 190
" " madreporianus, 190
Corbicula Saulcyi, 73, 205
" syriaca, 205
Corchorus trilocularis, 86, 159
Coriandrum sativum, 133
Corvus affinis, 40, 221, 228

Corvus corax, 15, 221
" cornix, 221
" frugilegus, 221
" umbrinus, 5, 15, 221
Cotile rupestris, 218
Cotula cinerea, 32, 96
Coturnix communis, 225
Cotyle obsoleta, 218, 228
" palustris, 228
" rupestris, 24
Cotyledon, 39
" umbilicus, 17, 19, 35, 93, 133
Cratægus aronia, 26
" azarolus, 64, 92
" sinaica, 17, 19, 92, 133
Crepis senecioides, 68, 98, 165
Cressa cretica, 9, 29, 101, 136
Critullus colocynthis, 5
Crotalaria ægyptiaca, 11, 89, 131
Crozophora obliqua, 12, 16, 110, 139, 162
" tinctoria, 139
Crucianella ciliata, 133
" membranacea, 133
Cuculus canorus, 229
Cucumis, 58
" prophetarum, 9, 29, 32, 93, 133, 160
Cucumis trigonus, 93
Cucurbitaceæ, 18
Cupressus pyramidalis, 70
" sempervirens, 18, 70
Cuscuta, 73
" arabica, 136
" palæstina, 73, 101
Cyanecula cæruleculus, 4, 29, 216
Cyclamen, 61
" latifolium, 59, 61, 68, 99
Cymodocea ciliata, 139
" isoetifolia, 139
" rotundata, 139
Cynanchum acutum, 42, 52, 100
Cynocrambe prostrata, 106
Cynodon dactylon, 9, 32, 117, 141
Cyperaceæ, 51
" eleusinoides, 51
" papyrus, 51
Cyperus conglomeratus, 140
" distachyus, 114, 140
" eleusinoides, 53, 114, 163, 167
" lævigatus, 23, 61, 114, 140, 167
" longus, 61, 115
" papyrus, 114
" rotundus, 64, 115
" schœnoides, 64, 115
Cypræa, 29
" erosa, 191

31—2

Cypræa gangrenosa, 192
,, globulus, 192
,, isabella, 191
,, (Trivia) oryza, 192
,, pulex, 197
,, scurra, 201
,, tabescens, 191
,, turdus, 191
Cyprinodon dispar, 54
Cypselus affinis, 229
Cyrtacanthacris notata, 183
,, ornatipes, 183
Cytheræa blanda, 201
,, florida, 194

D

Dæmia, 26, 35, 58
,, cordata, 12, 15, 24, 26, 100, 135, 161
Danthonia, 32
,, Forskahlii, 11, 116, 141, 167
Daphne acuminata, 38
,, linearifolia, 108, 167
,, mucronata, 38
Datura stramonium, 102
Dead Sea Basin, Tropical Flora of, 159
Dead Sea Valley, Flora of the, 156
Delta, Flora of the, 151
Description of Sinai, 123
Desert birds, 153
,, flora, 146
,, plants, 147
,, species, 146, 148, 153
Deverra, 6, 39
,, tortuosa, 5, 27, 62, 94, 133
,, triradiata, 133
Dianthus multipunctatus, 32, 37, 62, 83
,, sinaicus, 17, 83, 84, 129
Dicranella varia, 171
Digera arvensis, 53, 108, 162, 166
Diplodonta rotundata, 195, 200
Diplotaxis acris, 129
,, harra, 81, 129
,, pendula, 38, 81
Dipus ægypticus, 236
Dolichos lablab, 66, 91
Dorylus juvenculus, 180
Draba, 71
Drymœca, 22
,, gracilis, 217
,, inquieta, 4, 217, 228

E

Echaverias, 66
Echinops, 39
,, glaberrimus, 17, 19, 37, 97, 135, 165

Echinospermum sinaicum, 136
,, spinocarpos, 136
Echiochilon fruticosum, 12, 65, 136
Echium longifolium, 136
,, plantagineum, 63, 101
,, Rauwolfii, 136
Eclipta, 165
,, alba, 53, 96
Egretta gazetta, 229
Eleusine indica, 141
Elionurus (Cœlorachis) hirsuta, 10
,, hirsutus, 7, 116, 141
Emberiza miliaria, 220
,, striolata, 8, 220, 228
Emex spinosus, 64, 71, 108
Enarthrocarpus lyratus, 58, 59, 82, 163
,, strangulatus, 82, 129
Endocarpon hepaticum, 120, 172
Entosthodon Templetoni, 20, 119, 170
Ephedra, 39
,, alata, 65, 68, 142
,, alte, 19, 26, 35, 118, 142
,, elata, 118
,, fragilis, 118
Epilachna chrysomelina, 179
Equisetum elongatum, 118, 167
,, ramosum, 142
Eragrostis cynosuroides, 32, 51, 117
,, megastachya, 117, 167
,, pilosia, 117, 167
,, pœoides, 117, 142, 167
Eremias guttata, 8, 210
,, gutto-lineata, 4, 210
Eremobium lineare, 44, 80, 128
Eremostachys laciniata, 59, 105
Erianthus ravennæ, 51, 116
Erigeron Bovei, 24, 134, 160
,, (Conyza) Bovei, 96
,, trilobum, 134
Erinaceus brachydactylus, 237
Erithacus rubecula, 216
Erodium bryoniæfolium, 65, 87, 131
,, cicutarium, 59, 86, 131
,, glaucophyllum, 5, 87
,, gruinum, 68, 87
,, hirtum, 35, 62, 87, 131
,, laciniatum, 87, 131
,, malacoides, 68, 87
,, moschatum, 68, 86
Erophila vulgaris, 81
Erucaria, 71
,, aleppica, 42, 81, 128
,, microcarpa, 62, 81
Erygeron, 165
Eryngium, 32, 94
Erythræa ramosissima, 136

Erythræa spicata, 100, 136, 165
Erythrospiza githaginea, 12, 220, 228
Erythrostictus palæstinus, 140
Eucalypta vulgaris, 20, 119, 170
Eucalyptus, 66
Euphorbia, 71
,, ægyptiaca, 53, 109, 162, 167
,, aulacosperma, 68, 71, 110
,, chamæpeplus, 110, 139
,, cornuta, 9, 109, 139
,, exigua, 63, 110
,, granulata, 139, 162
,, helioscopia, 110, 139
,, obovata, 139
,, paralias, 110
,, peploides, 64, 110
,, peplus, 110, 139
,, terracina, 63, 110, 139
Evax anatolica, 96
,, contracta, 73, 96
Exocœtus, 28

F

Fagonia, 14, 18, 58
,, arabica, 87, 131
,, Bruguieri, 131
,, cretica, 5, 9, 18, 23, 87
,, glutinosa, 87, 131
,, grandiflora, 87
,, kahirica, 87
,, kahirina, 131
,, mollis, 131
,, myricantha, 14, 32, 39, 87, 131
Falco lanarius, 228
,, tinnunculus, 223
Farsetia ægyptiaca, 6, 25, 80, 128
,, ovalis, 128
Faunas of the Red Sea and Mediterranean, 199
Felis maniculata, 60, 237
,, pardus, 236
Ferns, 150
Ferrusacia thamnophila, 53, 204
Ferula sinaica, 133
Ficus carica, 42, 110
,, pseudosycomorus, 18, 139
,, sycomorus, 32, 40, 64, 110
Filago prostrata, 25, 96
Fimbristylis dichotoma, 52, 64, 115
,, ferruginea, 115
Fissurella Ruppellii, 195
Flora of the Ghôr es Safieh, 152
Flora of the Ghôr or Valley of the Dead Sea, 156

Flora of Sinai, Remarks on table of, 125
Flora of the Sinaitic Peninsula, Tabular view of the, 128-144
Flora, Spanish, 157
Floras, Lowland, 152
Forficula Lucasi, 182
Forskahlea, 22, 35, 58
,, declivis, 192
,, tenacissima, 110, 139
Fossombronia angulosa, 119, 170
Francoeuria crispa, 134
Frankenia hirsuta, 130
,, pulverulenta, 130
Fraxinus, 40
Fresh-water and Land Mollusca, 202
Fringilla cœlebs, 220
Fuligula ferina, 224
Funaria, 71
,, capreolata, 80
,, micrantha, 34, 80
,, parviflora, 80
Funaria calcarea, 171
,, hygrometrica, 119, 170, 171
Fusus polygonoides, 189

G

Gagea reticulata, 68, 113, 140
Galbella, 48
,, Beccari, 48
,, Harti, 178, 184
Galeodes araneoides, 176
Galerita arenicola, 229
,, cristata, 221
Galium, 71
,, aparine, 95
,, canum, 37, 94
,, Decaisnei, 133
,, judaicum, 68, 71, 95
,, petræ, 95, 164
,, sinaicum, 94, 133
,, spurium, 133
,, tricorne, 133
Gallinago media, 225
Garrulus atricapillus, 221
Gazella dorcas, 234
Gena lutea, 192
Genista (Retama) retam, 89
Geranium, 39
,, molle, 68, 86
,, tuberosum, 37, 39, 86
Gerbillus erythrurus, 44, 236
,, pygargus, 236
Ghôr, Flora of the, 156, 157
Glacial period, 150, 153, 157

Glaciers on Lebanon, 149
Glaucium arabicum, 14, 21, 32, 80, 128
Glinus lotoides, 11, 85, 130
Globularia, 39
„ alypum, 104
„ arabica, 34, 137
Glossonema Boveanum, 161
Gomphocarpus, 15, 18, 26, 66
„ sinaicus, 14, 15, 16, 18, 34, 100, 135
Gongylus ocellatus, 210
Grimmia apocarpa, 20, 118, 168
„ crinita, 118, 169
„ leucophæa, 20, 118, 169
„ pulvinata, 118, 169
„ trichophylla, 118, 169
Grus communis, 225
Gullium, 38
Gymnarrhena micrantha, 134
Gymnocarpum, 18
„ fruticosus, 7, 85, 130
Gymnocarpus fruticosus, 6, 24, 25
Gymnostomum calcareum, 171
„ rupestre, 20
„ verticillatum, 20
Gyps fulvus, 223
Gypsophila alpina, 20
„ elegans, 130
„ hirsuta, 20, 84, 130
„ Rokejeka, 14, 32, 84, 130

H

Hagioseris, 73, 98
Halcyon Smyrnensis, 222
Haliotis scutulum, 193
Halocnemum strobilaceum, 138
Halodule uninervis, 139
Halogeton alopecuroides, 138
Haloxylon articulatum, 138
Haplophila ovalis, 139
„ stipulacea, 139
Harpa minor, 191
„ solida, 200
Hedypnois cretica, 73, 98
„ galilæa, 73
Helianthemum kahiricum, 59, 83, 129
„ Lippii, 14, 26, 34, 83, 129
„ ventosum, 129
„ vesicarium, 129
Helices, 33
Heliotropium arbainense, 27, 29, 101, 136, 161
Heliotropium luteum, 6, 101, 136
„ persicum, 136
„ rotundifolium, 58, 101
„ undulatum, 14, 25, 101, 136

Helix bætica, 203
„ Boissieri, 61, 203
„ candidissima, 41, 61, 203
„ cavata, 61, 203
„ cæspitum, 41, 202
„ filia, 41, 203
„ hierochuntina, 202
„ joppensis, 61, 202
„ Ledereri, 61, 203
„ Olivieri, 203
„ prophetarum, 41, 203
„ protea, 61, 203
„ rustica, 218
„ Seetzeni, 41, 61, 203
„ spiriplana, 41, 203
„ syriaca, 61, 202
„ tuberculosa, 61, 203
„ vestalis, 61, 203
Helosciadium, 71
„ nodiflorum, 94
Henna (Lawsonia), 49
Herba alba, 17, 21
Herniaria cinerea, 130
„ (H. hemistemon), 20, 85, 130
Heterogamia maris-mortui, 182, 184
Hibiscus ovalifolius, 131
Himatismus villosus, 179
Hippobosca camelina, 182
Hippocrepis cornigera, 132
„ unisiliquosa, 71, 90
Hipponyx australis, 193
Hirundo Savignii, 218
„ urbica, 229
Holosteum liniflorum, 130
Holotermes vagans, 31, 182
Homalothecium sericeum, 171
Hussonia uncata, 128
Hyæna striata, 237
Hyalæa quadridentata, 189
„ uncinnata, 189, 199
Hyaloma ægyptium, 177
Hymenocarpus circinnatus, 71, 90
Hymenostylium rupestre, 169
„ (Gymnostomum) rupestre, 119
Hymenostylium (Eucladium) verticillatum, 119, 169
Hyoscyamus aureus, 23, 39, 68, 102
„ muticum, 23
„ muticus, 6, 43, 102, 137
„ pusillus, 137
Hypecoum procumbens, 71, 80
Hypericum sinaicum, 130
„ tetrapterum, 62, 86
Hyphæne thebaica, 9, 28, 111, 139
Hypnum aduncum, 171

INDEX

Hypnum (Brachythecium) Ehrenbergii, 119
,, Ehrenbergii, 170
,, ruscifolium, 20, 119, 170
,, rusciforme, 171
,, tenellum, 170
,, velutinum, 20, 119, 170
Hyrax syriacus, 21, 233
Hystrix cristata, 235
,, hirsutirostris, 235

I

Ianthina nitens, 193, 199
Ifloga, 35
,, spicata, 25, 96, 134
Imperata cylindrica, 22, 23, 51, 115, 141
Indigofera arabica, 132
,, argentea, 90
,, paucifolia, 52, 90, 160, 164
Insecta, etc., 175
Inula crithmoides, 52, 95
,, dysenterica, 95
,, maritimus, 114
,, viscosa, 39, 68, 95, 134
Iphiona juniperifolia, 13, 17, 33, 96, 134
,, montana, 14, 17
,, scabra, 22, 23, 32, 33, 96, 134
Iris (Xiphion) palæstinum, 112
Isatis, 81
,, aleppica, 81

J

Juncus acutus, 114
,, bufonius, 140
,, effusus, 140
,, inaritimus, 9, 29, 51, 52, 56, 140
,, paniculatus, 140
,, punctorius, 140
,, subulatus, 114
Juniperus, 117
,, phœnicea, 34, 38, 43, 117

K

Khardal, 49
Kochia latifolia, 138
,, muricata, 138
Koeleria phleoides, 73, 117, 142
,, sinaica, 142

L

Labiate, 19
Lacertidæ, 55
Lagoseris bifida, 135
Lakes, saline, 152

Lamarckia aurea, 71, 117
Lamium amplexicaule, 63, 105, 137
Land and Fresh-water Mollusca, 202
Lanicus collurio, 229
,, excubitor, 229
,, rufus, 229
Lanius fallax, 12, 29, 218, 228
Lantana camera, 3, 104
Larus gelastes, 229
,, melanocephalus, 230
,, minutus, 29, 226
,, ridibundus, 226
Lasiocampidæ, 182
Lasiopogon muscoides, 134
Lathyrus blepharicarpus, 68, 91
Latirus (Turbinella) turritus, 190
Lavandula, 25
,, coronopifolia, 12, 22, 104, 137
,, pubescens, 137
,, stœchas, 66, 104
Lawsonia alba, 66, 93
Lebanon, Glaciers on, 149, 151
,, Snow on, 151
,, species, Alpine, 149
Lecanora (Squamaria) crassa, 120, 172
,, (Sarcogyne) pruinosa, 120, 172
Ledum, 94
Leontice leontopetalum, 80
Leontodon arabicum, 135, 136
,, hispidulum, 135
Lepidium chalepense, 128
,, Draba, 128
Leptadenia pyrotechnica, 51, 58, 100, 136, 161
Lepturus incurvatus, 142
Lepus ægyptiacus, 30
,, ægyptius, 235
,, sinaiticus, 15, 234
Leucas inflata, 137
Leyssera capillifolia, 10, 96, 134
Lima hians, 193, 200
Limnæa peregra, 53, 204
Linaria ægyptiaca, 137
,, albifrons, 73, 103
,, elatine, 103
,, floribunda, 25, 102, 137
,, Hoelava, 64, 103
,, macilenta, 22, 23, 27, 29, 35, 103, 137, 165
,, micrantha, 73, 103
,, spuria, 137
Lindenbergia, 26
,, sinaica, 103, 137, 166
List of orders arranged according to the number of their Sinaitic species, 145
Lithospermum callosum, 9, 65, 101, 136

Lithospermum tenuiflorum, 17, 58, 101
Lolium multiflorum, 142
Loranthus, 55, 56
  ,, acaciæ, 31, 32, 42, 72, 109, 162
Lotononis dichotoma, 132
  ,, Leobordea, 13, 29, 89
  ,, persica, 132
Lotus arabicus, 132
  ,, lamprocarpus, 90
  ,, lanuginosus, 25, 90, 132
  ,, tenuifolius, 52, 90
Lucina dentifera, 194
  ,, semperiana, 195
Lunularia vulgaris, 120, 170
Lupinus reticulatus, 89
  ,, termis, 89
Lycium arabicum, 23, 43, 102, 162
  ,, europæum, 11, 12, 32, 35, 65, 102, 136
Lycopus europæus, 52, 104
Lygæus militaris, 183
Lysimachia dubia, 99
Lythrum, 71
  ,, hyssopifolium, 52, 93
  ,, salicaria, 52, 93

M

Machetes pugnax, 229
Mactra decora, 201
  ,, olorina, 195
  ,, semisulcata, 195
Malcolmia, 71
  ,, aculeolata, 128
  ,, africana, 128
  ,, crenulata, 68, 81
  ,, pulchella, 62, 68, 81, 128
Malva, 39, 71
  ,, parviflora, 131
  ,, rotundifolia, 12, 27, 86, 131
  ,, sylvestris, 86
Mandragora officinarum, 63, 102
Marginella, 195
Marrubium alysson, 59, 105
Matricaria aurea, 71, 96
  ,, auriculatum, 96
Matthiola arabica, 6, 128
  ,, humilis, 62, 80, 163
  ,, incana, 80
  ,, livida, 128
  ,, oxyceras, 73, 80, 128
Medicago ciliaris, 132
  ,, denticulata, 89
  ,, laciniata, 64, 90, 132
  ,, tribuloides, 132
Mediterranean Flora of Sinai, 149

Melania tuberculata, 4, 45, 204
Melanocorypha calandra, 222
Melanopsis buccinoidea, 45, 204
  ,, costata, 73, 205
  ,, eremita, 45, 205
  ,, procerosa, 205
  ,, Saulcyi, 45, 204
Melia azaderach, 65
Melilotus messanensis, 132
  ,, parviflora, 132
Mentha, 71
  ,, lavandulacea, 18
  ,, sylvestris, 104, 137
Mercurialis annua, 110
Mergus serrator, 229
Merops apiaster, 229
Mesembryanthemum Forskahlei, 133
Mesembryanthemums, 66
Mesotena punctipennis, 179
Micromeria barbata, 68
  ,, juliana, 104
  ,, myrtifolia, 25, 68
  ,, nervosa, 68, 104
  ,, serpyllifolia, 104
  ,, sinaica, 104, 137, 166
Microrhynchus, 58
  ,, nudicaulis, 32
Midian plants, 158
Milvus ægyptius, 228
  ,, nigrans, 228
  ,, regalis, 223
Mimosa, 66
Miocene times, 157
Mitra, 29
  ,, abbatis, 190
  ,, consanguinea, 190
  ,, corniculato, 195
  ,, cucumerina, 100
  ,, digitalis, 190
  ,, ebenus, 195
  ,, filosa, 190
  ,, modesta, 190
  ,, osiridis, 190
  ,, variegata, 190
Modiolo auriculata, 194
Modulus tectum, 192
Mœrua uniflora, 129
Mollusca fluviatile, 55
  ,, from 'Akabah, Red Sea, 189
Monodonta australis, 192
  ,, obscura, 192
Monsonia nivea, 11, 32, 44, 87, 131
Morettia, 35
  ,, canescens, 11, 80, 128
  ,, philæana, 128
Moricandia arvensis, 128

Moricandia clavata, 129
,, dumosa, 22, 23, 33 34, 37, 39, 81, 129
,, sinaica, 23, 81, 129
Moringa aptera, 50, 131, 160
Mosses, Palestine, 171
,, Sinaitic, 150, 171
Motacilla alba, 4, 15, 29, 217
,, flava, 29, 217
,, Rayi, 229
,, sulphurea, 218
Mount Hor, plants, 153
Murex anguliferus, 189
,, cristatus, 197
,, inflatus, 200
,, tribulus, 194
Mus bactrianus, 235
Muscari racemosum, 71, 113
Myoxus quercinus, 235
Mytilus variabilis, 195

N

Narcissus Tazettæ, 65, 112
Nassa coronata, 200
,, Cuvieri, 196
,, delicata, 190
,, gemmulata, 190
,, incrassata, 196
,, plebecula, 190
,, sordida, 190
Nasturtiopsis arabica, 128
Nasturtium, 71
,, officinale, 68, 71 80, 128
Natica Josephinia, 197
,, mamilla, 191
,, marochiensis, 191
,, melanostoma, 191
Neophron percnopterus, 5, 233
Nepeta septem-crenata, 18, 105, 137
Nerita albicilla, 195
,, antiquata, 194
,, marmorata, 195
,, polita, 192
,, Rumphii, 192
,, stella, 195
Neritina fluviatilis, 205
,, Michouxii, 205
Nerium oleander, 34, 99
Neslia paniculata, 71, 81
Neurada procumbens, 14, 15, 64, 92, 133
Nigella deserti, 128
Nitraria, 22, 88, 131
,, tridentata, 4, 16, 29
Noæa, 39
,, spinosissima, 39, 58, 107

Noctuidæ, 182
Nothochlœna lanuginosa, 142
Notholæna lanuginosa, 34, 118
Notoceras canariense, 58, 81, 128
Nubian and Abyssinian plants, 150
Nubk, Groves of, 153

O

Ochradenus, 22
,, baccatus, 8, 32, 35, 58, 71, 83, 129
Ocnera hispida, 179
Odontospermum graveolens, 5, 22
,, pygmæus, 32
Œdicnemus scolopax, 225
Œdipoda cærulans, 183
,, insubrica, 183
Oldenlandia Schimperi, 133
Olea europea, 99
Oligomeris subulata, 129
Omphalaria, 120, 172
Omphalius cariniferus, 192
Onobrychis Ptolemaica, 29, 91, 132
Ononis, 40
,, antiquorum, 73, 89
,, campestris, 89, 164
,, natrix, 64, 68, 89
,, serrata, 62, 89
,, sicula, 132
,, vaginalis, 38, 89
Onopodon ambiguum, 135
Onosma, 39
,, giganteum, 101
,, syriaca, 68
Opuntra vulgaris, 63
Origanum, 39
,, maru, 18, 23, 37, 104, 137
,, maru, ? forma, 104
Ornithogalum umbillatum, 60, 113
Orobanche ægyptiaca, 73
,, (Phelipæa) ægyptica, 103
,, cernua, 103
Oryctes nasicornis, 178
Osher (Calotropis), 49
Ostræa frons, 193
Otiona Aitonia, 119, 170
Otostegia Schimperi, 137
,, moluccoides, 137
Oxystelma alpina, 161

P

Pachdema sinaitica, 178
Palestine Flora, Additions to, 163

Palestine mosses, 171
Pancratium Sickembergeri, 11, 13, 16, 44, 112, 140
Panicum, 29
 " barbinode, 115, 167
 " colonum, 115
 " molle, 115, 167
 " repens, 115
 " teneriffæ, 25, 29, 115, 140
 " turgidum, 23, 44, 115, 140
Papaver Decaisnei, 128
Paphia glabrata, 194
Pappophorum brachstachyum, 142
Paracaryum micranthum, 136
 " rugulosum, 136
Parietaria alsinefolia, 139
 " judaica, 110
Paronychia argentea, 37, 59, 85
 " capitata, 85
 " desertorum, 6, 85, 130, 164
 " nivea, 63, 85, 164
 " sinaica, 130
Parus major, 217
Passer domesticus, 219
 " hispaniolensis, 25, 29, 219, 228
 " moabiticus, 219
Patella variabilis, 193
Pecten lividus, 193
 " pes-felis, 200, 201
 " sanguinolentus, 193
 " varius, 200
Pectunculus, 29
 " arabica, 195
 " glycimeris, 197
 " lineatus, 193
 " paucipictus, 201
 " pectiniformis, 194
 " siculus, 193, 200
Peganum, 27
 " harmala, 14, 19, 62, 88, 131
Pelecanus crispus, 223
Pennisetum cenchroides, 34, 39, 64, 115
 " ciliare, 141
 " dichotomum, 7, 23, 115, 141, 167
Pennisetum elatum, 141
 " orientale, 141
Pentatropis spiralis, 34, 53, 100, 161, 165
Periploca aphylla, 135
Perrinea stellata, 192
Persia plants, 150
Phæopappus scoparius, 135
Phagnalon nitidum, 16, 96, 134
Phalacrocorax carbo, 229
Phalaris minor, 73, 116, 141
Phasianella jaspidea, 192

Phasmoptynx capensis, 229
Phelipæa tubulosa, 137
Phlomis aurea, 18, 35, 37, 105, 137, 166
Phœnix dactylifera, 7, 18, 111, 140
Phorus corrugatus, 192
Phos roseatus, 190
Phragmites communis, 9, 117, 142
 " gigantea, 24, 117
Phylloscopus rufus, 4, 217
 " trochilus, 217
Physa cortorta, 53, 204
Picridium tingitanum, 135
Picris, 73
 " cyanocarpa, 135
Picus syriacus, 222
Pieris (Belenois) lordacea, 181
 " mesentina, 181
Pimpinella cretica, 68, 94, 133
Pinna, 193
Pinus halepensis, 67, 70, 117
Piptatherum multiflorum, 17, 116, 141
Pistacia, 39
 " palæstina, 37, 38, 68, 88
 " terebinthus, 131
Pisum fulvum, 68, 91
Planaxis Savignyi, 193
Planorbis albus, 53, 204
 " piscinarum, 204
Plantago albicans, 64, 106, 138
 " amplexicaulis, 138
 " arabica, 18, 106, 138
 " arenaria, 106
 " ciliata, 138
 " cylindrica, 138
 " lagopus, 68, 106
 " Loefflingii, 58, 106, 166
 " ovata, 25, 106, 138
 " phæostoma, 138
Plants, flowering, 156
Plateaux, or Montane, Flora of Sinai, 149
Pleurotoma cingulifera, 189
Plicatula ramosa, 195
Pliocene times, 150
Pluchea dioscorides, 42, 96
Poa annua, 63, 117
 " sinaica, 14, 17, 19, 117, 142
 " soongarica, 142
Podonosma syriaca, 34, 101
Pœcilocera bufonia, 183
Polistes, 180
Pollia bicolor, 196
 " Orbignyi, 196
Polycarpæa fragilis, 9, 85, 130
 " prostrata, 9, 26
 " (Robbairea) prostrata, 85

Polycarpon arabicum, 130
" succulentum, 62, 85, 130, 164
Polygala spinescens, 129
Polygonum equisetiforme, 39, 108
Polypogon maritimum, 141
" monspeliense, 116, 141
Polyrhachis seminiger, 53, 180
Populus euphratica, 26, 42, 44, 51, 72, 111, 139
Portulaca oleracea, 130
Poterium spinosum, 39, 92
" verrucosum, 34, 92, 133
Prassium majus, 64, 105
Pratincola rubicola, 216
Primula Boveana, 14, 17, 99, 135
Prionotheca coronata, 8, 179
Prosopis spicigera, 160
" stephaniana, 44, 64, 71, 72, 91
Prosthesina, 177
Psammomys, 44
" obesus, 60, 236
Psoralea, 14, 17
" bituminosa, 90, 132
" plicata, 14, 90
Pteranthus echinatus, 85, 130
Pterocephalus sanctus, 20, 37, 95, 134, 165
Pteroceras truncatus, 194
Pterocles coronatus, 228, 229
" senegalensis, 46
" senegalus, 224
Ptyodactylus Hasselquistii, 210
Pulicaria arabica, 44, 54, 95
" crispa, 18
" (Francoeuria) crispa, 23, 95
" longifolia, 134
" undulata, 13, 54, 95, 134, 161
Pupa Saulcyi, 204
Purpura serta, 190
Pycnocycla tomentosa, 133
Pycnonotus xanthopygus, 29, 72, 218
Pyrameis cardui, 15, 181
" Kershawi, 181
Pyrethrum auriculatum, 134
" santalinoides, 17, 96, 134
Pyrrhocoris apterus, 183
Pyrula paradisiaca, 195

Q

Quercus coccifera, 43, 67, 110

R

Radula inflata, 200
Rallus aquaticus, 225
Ramalina crispatula, 120, 172

Ramnus palæstina, 131
Rana esculenta, 210
Ranella (Lampas) affinis, 190
" granifera, 190
" rhodostoma, 190
Ranunculus asiaticus, 73, 79
Raphanus sativus, 129
Reaumuria, 6, 18, 25
" hirtella, 86, 130
" palæstina, 86
" vermicularis, 5, 35
" vermiculata, 86
Red Sea, shells of the, 199
" species from the, 201
Reptilia, 55
Reseda alba, 68, 83
" amblyocarpa, 83
" arabica, 129
" muricata, 129
" propinqua, 129
" pruinosa, 23, 25, 71, 83, 129
" stenostachya, 83, 129
Resedaceæ, 58
Retem, 26
Retama, 22, 35, 58
" retam, 5, 17, 65, 66, 71, 132
Rhamnus, 39
" oleoides, 44, 50, 89
" punctata, 37, 39, 43, 66, 88, 89
Rhus oxyacanthoides, 43, 73, 88
Rhyncostegium tenellum, 171
Rhynchœa capensis, 61, 225
Rhynchosia minima, 52, 91, 160, 164
Rhynchostegium (Hypnum) rusciforme, 171
Rhyncocalamus melanocephalus, 41, 209
Riccia lamellosa, 59, 120, 171
Ricinula (Pentadactylus) ablolabris, 190
" " fiscillum, 190
" Morio, 201
" (Pentadactylus) spectrum, 190
" (Purpura) tuberculata, 190
Ricinus communis, 51, 64, 66, 110
Ridolfia segetum, 133
Robbairea prostrata, 130
Rœmeria orientalis, 128
Rosa rubiginosa, 133
Rubia Olivieri, 65, 94
" peregrina, 39, 94, 164
Rubus, 71
" discolor, 68, 92
Rumex dentatus, 139
" obtusifolius, 65, 108
" roseus, 39, 108, 139
" vesicarius, 108, 139
Ruppia, 51, 52
" rostellata, 139

32—2

Ruppia spiralis, 51, 52
Ruta, 27
" graveolens, 65, 88
" tuberculata, 13, 18, 19, 21, 27, 35, 131
Ruta (Haplophyllum) tuberculatum, 88
Ruticilla phœnicura, 229
" tithys, 216

S

Saccharum ægyptiacum, 42, 116, 141
Sahadan, 28
Sahara a sea-bed, 149
Salicornia, 51
" fruticosa, 106
" herbacea, 51, 52, 106
Saline lakes, 152
Salix acmophylla, 42, 110, 167
" babylonica, 139
" safsaf, 18, 20, 26, 27, 139
Salsola fœtida, 32, 44, 52, 58, 107, 138, 162, 166
Salsola inermis, 39, 58, 64, 107, 166
" longifolia, 107, 166
" rigida, 39, 58, 107
" tenuifolia, 58
" tetragona, 42, 51, 58, 107, 138
Salsolaceæ, 47, 53, 58
Salvadora (Arak), 49
" persica, 99, 135, 161
Salvia ægyptiaca, 34, 59, 105, 137
" controversa, 59, 105, 137
" deserti, 29, 105, 137, 166
" palæstina, 137
" triloba, 67, 105
" verbenaca, 59, 105
" viridis, 105
Santolina fragrantissima, 14, 16, 22, 97, 134
Saponaria vaccaria, 73, 83, 130
Saturela cunefolia, 32, 43, 59
" " forma, 104
Savignya ægyptiaca, 129
Saxicola amphileuca, 229
" deserti, 227
" eurymelæna, 229
" Finschii, 68, 216
" isabellina, 8, 215
" leucocephala, 216, 227
" leucopygia, 12, 21, 216
" lugens, 21, 68, 72, 216, 227
" mœsta, 215
" monacha, 10, 215, 227
" œnanthe, 216
Scabiosa eremophila, 134
Scandix pinnatifida, 133

Schanginia baccata, 138
Schimpera arabica, 128
Schismus arabicus, 142
" marginatus, 73, 117
Schœnus nigricans, 140
Schouwia Schimperi, 129
Scilla, 39
Scirpus holoschœnus, 19, 115, 140
" maritimus, 52, 115, 140
Sclerocephalus arabicus, 52, 85, 130, 159, 164
Scleron orientale, 179
Scleropoa memphitica, 142
Scolopendra canidens. 177
" centipede, 41
Scorzonera alexandrina, 62, 98, 165
" lanata, 59, 98
Scrophulariaceæ, 19, 38
Scrophularia canina, 103
" deserti, 23, 32, 103. 137
" heterophylla, 37, 103, 166
" variegata, 103, 137
" zanthoglossa, 63, 103
Sedum altissimum, 68
Seetzenia orientalis, 11, 88, 131
Senecio coronopifolius, 59, 97, 135
" Decaisnei, 134
" vernalis, 97
Serinus meridionalis, 229
Sesamum indicum, 70
Setaria glauca, 140
" verticillata, 140
Shaphan (the daman or wabar of the Arabs), 233
Sherardia arvensis, 68, 94
Silene ægyptiaca, 84
" arabica, 130
" atocion, 84
" colorata, 84, 163
" conoidea, 130
" dichtoma, 59, 84
" Hussoni, 59, 163, 184
" inflata, 65, 84
" leucophylla, 130
" linearis, 130
" palæstina, 73, 84
" rigidula, 62
" Schimperiana, 130
" succulenta, 63, 84
" villosa, 130
Sinai, description of, 123
" dried-up lakes in, 150
" Mediterranean flora of, 149
" Plateaux, or Montane, Flora of, 149
" Vegetation in, 151
Sinaitic flora, Distribution of, 154

Sinaitic mosses, 171
" Peninsula, Analysis of orders represented in the, 143, 144
" species, 145, 154
Sinapis alba, 82
" arvensis, 82, 129
Siphonaria cochleariformis, 193
" lecanium, 193
Sisymbrium erysimoides, 80, 128, 163
" irio, 81, 128
" pannonicum, 128
" Schimperi, 128
Solanum, 71
" coagulans, 102, 162
" nigrum, 32, 102, 136
Solarium cingulum, 191
" perspectivum, 201
Solenostoma argel, 135, 161
Sonchus maritimus, 52, 98, 165
" nudicaulis, 26
" (Microrhynchus) nudicaulis, 23
" oleraceus, 135
" spinosus, 13, 22, 29
Sorghum, 70
" halepense, 116
Sources of information, 125
Spalax typhlus, 60, 236
Spanish flora, 157
Sparacis, 177
Spergula pentandra, 84
Spergularia diandra, 130
" marginata, 61, 84
Sphænops capistratus, 8, 210
Spirostreptus bottæ, 177
" millipede, 41
Spondylus, 193
Sporobolus spicatus, 4, 116, 141, 167
Stachys, 22
" affinis, 14, 22, 105, 137
Statice pruinosa, 29, 44, 105, 138
" Thouini, 105, 138
Stellaria media, 84
Stenodactylus guttatus, 210
Stenopteryx hybridalis, 182
Steraspis squamosa, 178
Sterna minuta, 226
Sternbergia macrantha, 37, 39, 40, 112
Stipa barbata, 141
Strombus, 29
" auris-dianæ, 191
" floridus, 191
" fusiformis, 191
" gibberulus, 191
" lineatus, 194
" plicatus, 191
" tricornis, 194

Sturnus unicolor, 220
" vulgaris, 220
Suæda asphaltica, 44, 106
" fruticosa, 107
" monoica, 17, 22, 106, 138, 162
" vermiculata, 9, 138
Suædas, 51
Sub-fossil shells, 150
Subemarginula imbricata, 192
Succinea elegans, 202
Sus scrofa, 233
Sylvia Bonellii, 229
" deserti, 228
" melanocephala, 229
" nana, 22, 216, 228
" Ruepellii, 229
" sarda, 228

T

Table of Flora of Sinai, Remarks on, 125
Tabular view of the Flora of the Sinaitic Peninsula, 128-144
Tamarindus indica, 28
Tamarisk, 32
" articulata, 65
" Groves of, 153
Tamarix, 22
" articulata, 4, 86, 130, 164
" gallica, 71, 85, 130
" mannifera, 86
" nilotica, 4, 7, 15, 16, 22, 85
Tapes geographica, 197
Tellina scobinata, 194
" staurella, 201
Telephium sphærospermum, 130
Telphusa fluviatilis, 211
" (Potamophilon) fluviatilis, 56
Tephrosia, 90
" apollinea, 29, 90, 132
" purpurea, 23, 24, 29, 132
Teracolus chrysonome, 181
" phisadia, 181
Terebra affinis, 191
" babylonica, 191
" cærulescens, 200
" crenulata, 200
" dimidiata, 200
" nubeculata, 200
Tetrapogon villosum, 141
Teucrium leucocladum, 138
" polium, 14, 68, 105
" sinaicum, 17, 18, 23, 37, 105, 138, 166
Thelygonum cynocrambe, 68, 106
Thlaspi, 71

Thlaspi perfoliatum, 81
," perfoliatus, 68, 71
Thrincia tuberosa, 62, 98
Thymelæa, 40
," hirsuta, 39, 109, 139
," passerina, 39, 40
Thymus capitatus, 66, 104
," decussatus, 137
Tolpis altissima, 62, 98
Tordylium brachycarpa, 68
Torilis leptophylla, 68
," nodosa, 68, 94
," trichosperma, 68
Tortula, 35
," ambigua, 119, 169
," inermis, 20, 119, 169
," membranifolia, 119, 169
," muralis, 59, 169
," nitida, 119, 169
," revoluta, 119, 169
," rigida, 170
," (Trichostomum) rigidula, 119, 169
," ruralis, 119
," tophacea, 119, 169
," unguiculata, 119, 169
," vinealis, 119, 169
Totanus calidris, 226
," glottis, 229
," ochropus, 226
Traganum nudatum, 32, 107, 138
Trianthema pentandra, 53, 93, 164
," pentandrum, 160
Tribulus alatus, 25, 87, 131
," bimucronatus, 131
," terrestris, 14, 42, 54, 87, 131
Trichodesma, 32, 35
," africana, 14
," africanum, 29, 102, 136, 161
Trichostomum aviculare, 171
," barbula, 171
," nitidum, 171
Tridacna, 29
," crocea, 193
," elongata, 193
," gigas, 201
Trigonella arabica, 71, 89, 132
," fœnum-græcum, 132
," stellata, 132
Tringa alpina, 226
," minuta, 229
Tringoides hypoleucus, 29, 226
Tripteris Vaillantii, 97, 135, 165
Trisetum lineare, 141
," pumilum, 141
Tristram's Grakle, 40
Triton, 29

Triton anus, 190
," aquatilus, 190
," lampas, 189
," maculosus, 201
," pilearis, 189
," rubecula, 189
," trilineatus, 189
Trochus adriatica, 197
," dentatus, 192
," erythræus, 192
," sanguinolentus, 201
Tropical Flora of the Dead Sea Basin, 159
Tryxalis unguiculata, 17, 183
Turbo margaritaceus, 192
," sp. (?), 192
Turdus iliacus, 73, 215
," merula, 215
Turritella trisulcata, 192
Turtor risorius, 54, 224
Turtus senegalensis, 54, 224
Typha angustata, 9, 23, 24, 111, 140, 167

U

Umbilicus intermedius, 93
Upupa epops, 229
Urginea scilla, 32, 37, 58, 60, 62, 113, 233
," undulata, 44, 59, 61, 113, 167, 233
Uropetalum erythræum, 13, 113, 140
Urospermum picroides, 68, 98, 110, 135
Urtica pilulifera, 110
," urens, 110

V

Vanellus cristatus, 225
," gregarius, 229
Varthamia montana, 33, 37, 96, 134, 165
Venus, 29
," costellifera, 194
," crispata, 201
," reticulata, 194
," toreuma, 194
Verbascum Schimperianum, 137
," sinaiticum, 18, 19, 22, 25, 40, 102, 137
Verbascum sinuatum, 25, 39, 54, 102, 137
Verbena, 71, 74
," officinalis, 104
Veronica, 52, 71
," anagallis, 68, 103, 137
," beccabunga, 69, 103, 137
," camplopoda, 137
," syriaca, 20
Vespa orientalis, 9, 180
Vicia palæstina, 91, 103
," sativa, 91

Viscum cruciatum, 68, 70, 109
Vitex agnus-castus, 73, 74, 104
Volvula acuminata, 200
Vulpes nilotica, 60, 237
Vulpia myuros, 142
Vulsella spongiarum, 195

W

Withania somnifera, 63, 102, 136

X

Xanthium strumarium, 64, 96
Xiphion palæstinum, 60

Z

Zamenis atrovirens, 10, 60, 209
,, carbonarius, 209
,, Cliffordii, 41, 210
,, elegantissimus, 28, 209
,, ventrimaculatus, 21, 25, 28, 209
Zilla, 22
,, myagroides, 5, 16, 22, 26, 58, 82, 129
,, nitraria, 7, 35

Zitani, 28
Zizyphus, 18, 25, 42, 50
,, (Henna) 28, 30
,, spina-christi, 12, 72, 79, 88, 131
Zollikoferia, 39, 58, 98
,, arabica, 135
,, casinianæ, 39, 165
,, fallax, 135
,. glomerata, 135
,, massavensis, 135
,, mucronata, 135
,, nudicaulis, 98, 135
,, spinosa, 98, 135
,, stenocephala, 58, 98, 165
,, tenuiloba, 135
Zozimia absynthifolia, 133
Zukkum (Balanites), 49, 50
Zygophylla, 58
Zygophyllum, 18
,, album, 9, 16, 26, 71, 88, 131
,, coccineum, 23, 35, 88, 131
,, decumbens, 131
,, dumosum, 25, 26, 87, 131
,, simplex, 52, 58, 88, 131, 164

THE END.

BILLING AND SONS, PRINTERS, GUILDFORD.

www.ingramcontent.com/pod-product-compliance
Lightning Source LLC
Chambersburg PA
CBHW031336230426
43670CB00006B/347